P9-EMH-869

The Scientific Sherlock Holmes

The Scientific Sherlock Holmes

CRACKING THE CASE WITH SCIENCE AND FORENSICS

JAMES F. O'BRIEN

OXFORD
UNIVERSITY PRESS

Oxford University Press is a department of the University of Oxford.
It furthers the University's objective of excellence in research, scholarship,
and education by publishing worldwide.

Oxford New York
Auckland Cape Town Dar es Salaam Hong Kong Karachi
Kuala Lumpur Madrid Melbourne Mexico City Nairobi
New Delhi Shanghai Taipei Toronto

With offices in
Argentina Austria Brazil Chile Czech Republic France Greece
Guatemala Hungary Italy Japan Poland Portugal Singapore
South Korea Switzerland Thailand Turkey Ukraine Vietnam

Published in the United States of America by
Oxford University Press
198 Madison Avenue, New York, NY 10016

Library of Congress Cataloging-in-Publication Data
O'Brien, James F., 1941–
The scientific Sherlock Holmes : cracking the case with science and forensics /
James F. O'Brien.
p. cm.
Includes bibliographical references and index.
ISBN 978–0–19–979496–6 (hardcover)
1. Forensic sciences—History. 2. Chemistry, Forensic—History. 3.
Criminal investigation—History. 4. Detective and mystery stories, English—
History and criticism. 5. Holmes, Sherlock (Fictitious character)
6. Science in literature. I. Title.
HV8073.O36 2013
363.25—dc23
2012023297

ISBN 978–0–19–979496–6

9 8 7 6 5 4 3 2 1
Printed in the United States of America
on acid-free paper

To Ted, who would have loved this book.

Contents

Preface

Few characters in literature are more universally recognized than Sherlock Holmes. The subject of sixty stories by Arthur Conan Doyle and countless pastiches by other authors (not to mention even a "biography" or two), Holmes is nothing short of an icon of literature. While readers are captivated by his powers of observation and deductive reasoning, somewhat overlooked in the stories is the use of science and forensic methods, long before network television made them so popular. Conan Doyle (and Holmes) blazed a new trail in this regard, adding depth and complexity to the detective genre started by Edgar Allan Poe. This book focuses on the scientific aspects of Sherlock Holmes. Essentially every one of the sixty stories has some mention of science. In some of the stories, science is the dominant factor.

We begin by tracing the origins of Arthur Conan Doyle's science-oriented detective. Then, after describing the main characters in the stories in chapter 2, chapter 3 takes a detailed look at how Holmes used science to solve his cases. Because Sherlock Holmes knows more chemistry than any other science, chapter 4 examines Holmes the chemist. The final chapter looks at his knowledge and use of other sciences. Throughout the book, we use the terms "Sherlockian" and "Holmesian"[1] interchangeably to refer to someone with great interest and/or expertise in Sherlock Holmes.

[1] Sherlockian tends to be used in the United States and Holmesian in the United Kingdom (King, L. R., in King and Klinger 2011).

Acknowledgments

I wish to acknowledge the yeoman work done by my son, Mike O'Brien, in reading the entire manuscript and providing so much useful advice. I am grateful to Lorraine Sandstrom, Trint Williams, Sarah Pearl, and Rich Biagioni for help with some of the figures. Early discussions with my brother, Tom O'Brien, were much appreciated. Thank you to my editors, Jeremy Lewis and Hallie Stebbins. Maria Pucci helped me overcome my deficiencies with computers.

Finally I thank my wife, Barbara O'Brien, for reading the manuscript and providing crucial input on how the material should flow in its presentation. She also tolerated the mess in the family room for more than a year.

Introduction

Sherlock Holmes is the most recognizable character in all of literature. The first Sherlock Holmes story, *A Study in Scarlet* (STUD), was published in 1887. Today, over 125 years later, when a deerstalker hat is seen in a book, movie, TV ad, or billboard, the public automatically thinks "Sherlock Holmes." Old movies run on television again and again. New movies are made with consistent regularity. Plays are done all around the country and the world. Respectable presses publish Sherlock Holmes journals. There are even several Sherlock Holmes encyclopedias (Tracy 1977; Bunson 1994; Park 1994). While limited to sixty original stories by Arthur Conan Doyle, Sherlock Holmes buffs eagerly seek out new Holmes stories by would-be Conan Doyles. They call such stories "pastiches" and are easy marks for even marginal literature. Aspiring authors frequently base their stories on one of the more than one hundred cases mentioned by Doyle but not reported in full (Redmond 1982, xv; Jones 2011). Of course, "stories about the stories" are also coveted. Numerous Holmes societies exist in the United States and around the world. In the United States, the pinnacle of achievement for a Sherlock Holmes buff is an invitation to be a "Baker Street Irregular," a group apparently as odd as Holmes's ragamuffin street urchins from whom it takes its name.

Why is all this so? One reason for Holmes's appeal is that he is a flawed character. For instance, contrary to his image, he does not always correctly solve his cases. He admits that he failed four times. When reading a Holmes story, the reader can't be sure he will solve it, for even the master detective sometimes fails. Another flaw is his well-known drug dependence, which is discussed later.

Also among the primary reasons for the enduring popularity of Sherlock Holmes is his ability to make brilliant deductions. Readers continue to be fascinated by the way he can reason his way to the correct solution. In the opening of the first story, STUD, Holmes's first words to Dr. Watson are "How are you? You have been in Afghanistan, I perceive." Watson thinks someone has told Holmes this fact. But Holmes later explains how he deduced it from the doctor's appearance. *Shoscombe Old Place* (SHOS) is the sixtieth and last Sherlock Holmes story, published in 1927. In it, the fact that Sir Robert Norberton has given away his sister's beloved spaniel puzzles everyone but Holmes. The absence of the dog allows Holmes to deduce that the sister, Lady Beatrice Falder, has died and that Sir Robert is concealing that fact. Immediately everything makes sense and the case becomes easy for Holmes to solve. Whatever else changed in the Sherlock Holmes stories, Conan Doyle kept Holmes deducing throughout the entire forty years from 1887 to 1927.

In this study, we suggest that another strong component of the character's ongoing appeal and success is his knowledge of science and frequent use of the scientific method. Doyle himself, in an article in *Tit-Bits* on December 15, 1900, described how he tried to make his detective stories more realistic than the ones he had been reading (Green 1983, 346):

> I had been reading some detective stories, and it struck me what nonsense they were, to put it mildly, because for getting the solution to the mystery the authors always depended on some coincidence.

So he resolved to diminish the role of chance by having his detective employ science and reasoning on his way to the answer. With Poe's Dupin in mind, Doyle set out to make Holmes somewhat different. He tells us:

> Where Holmes differed from Dupin was that he had an immense fund of exact knowledge to draw upon in consequence of his previous scientific education.

Sherlock Holmes's knowledge of science not only provides fodder for debate among the legions of fans, it also lends credibility to his impressive powers of reasoning. Indeed, among the best-loved stories involving the detective, those that rely not just on deductive reasoning but also employ elements of science are regarded the most highly.

This book focuses on the scientific side of Sherlock Holmes. Initially we look at how the Holmes Canon came to be written. Chapter 2 introduces the main characters: Holmes, Dr. Watson, Professor Moriarty, and Holmes's brilliant brother, Mycroft. In chapter 3 we examine how Sherlock Holmes used scientific forensic techniques in his investigations. Chapters 4 and 5 deal with all of the science that was not used to solve crimes. Chapter 4 describes the chemistry that permeates the entire Canon. Chapter 5 deals with six other sciences that come up in the stories. Finally we conclude with some closing thoughts on Holmes's use of science and its contribution to the enduring appeal of the stories.

Codes

Much of the Holmesian world uses the following four-letter abbreviations for the names of the sixty stories written by Sir Arthur Conan Doyle. We use them extensively to avoid constant repetition of the titles. Also for brevity, the words "The Adventure of" are deleted from the many titles that contain them. Collectively the Sherlock Holmes stories are sometimes affectionately referred to as the "Canon."

Code	Title	Publication Date	Story #
ABBE	The Abbey Grange	Sept. 1904	39
BERY	The Beryl Coronet	May 1892	13
BLAC	Black Peter	Feb. 1904	33
BLAN	The Blanched Soldier	Oct. 1926	56
BLUE	The Blue Carbuncle	Jan. 1892	9
BOSC	The Boscombe Valley Mystery	Oct. 1891	6
BRUC	The Bruce-Partington Plans	Dec. 1908	42
CARD	The Cardboard Box	Jan. 1893	16
CHAS	Charles Augustus Milverton	March 1904	34
COPP	The Copper Beeches	June 1892	14
CREE	The Creeping Man	March 1923	51
CROO	The Crooked Man	July 1893	22
DANC	The Dancing Men	Dec. 1903	30
DEVI	The Devil's Foot	Dec. 1910	43

DYIN	The Dying Detective	Nov. 1913	46
EMPT	The Empty House	Sept. 1903	28
ENGR	The Engineer's Thumb	March 1892	11
FINA	The Final Problem	Dec. 1893	26
FIVE	The Five Orange Pips	Nov. 1891	7
GLOR	The "Gloria Scott"	April 1893	19
GOLD	The Golden Pince-Nez	July 1904	37
GREE	The Greek Interpreter	Sept. 1893	24
HOUN	The Hound of the Baskervilles	Aug. 1901	27
IDEN	A Case of Identity	Sept. 1891	5
ILLU	The Illustrious Client	Nov. 1924	54
LADY	The Disappearance of Lady Francis Carfax	Dec. 1911	45
LAST	His Last Bow	Sept. 1917	48
LION	The Lion's Mane	Nov. 1926	57
MAZA	The Mazarin Stone	Oct. 1921	49
MISS	The Missing Three-Quarter	Aug. 1904	38
MUSG	The Musgrave Ritual	May 1893	20
NAVA	The Naval Treaty	Oct. 1893	25
NOBL	The Noble Bachelor	April 1892	12
NORW	The Norwood Builder	Oct. 1903	29
PRIO	The Priory School	Jan. 1904	32
REDC	The Red Circle	March 1911	44
REDH	The Red-Headed League	Aug. 1891	4
REIG	The Reigate Squires	June 1893	21
RESI	The Resident Patient	Aug. 1893	23
RETI	The Retired Colourman	Dec. 1926	58
SCAN	A Scandal in Bohemia	July 1891	3
SECO	The Second Stain	Dec. 1904	40
SHOS	Shoscombe Old Place	March 1927	60
SIGN	The Sign of Four	Feb. 1890	2
SILV	Silver Blaze	Dec. 1892	15

SIXN	The Six Napoleons	April 1904	35
SOLI	The Solitary Cyclist	Dec. 1903	31
SPEC	The Speckled Band	Feb. 1892	10
STOC	The Stock-broker's Clerk	March 1893	18
STUD	A Study in Scarlet	Nov. 1887	1
SUSS	The Sussex Vampire	Jan. 1924	52
THOR	The Problem of Thor Bridge	Feb. 1922	50
3GAB	The Three Gables	Sept. 1926	55
3GAR	The Three Garridebs	Oct. 1924	53
3STU	The Three Students	June 1904	36
TWIS	The Man With the Twisted Lip	Dec. 1891	8
VALL	The Valley of Fear	Sept. 1914	47
VEIL	The Veiled Lodger	Jan. 1927	59
WIST	Wisteria Lodge	Aug. 1908	41
YELL	The Yellow Face	Feb. 1893	17

The Sherlockian Canon

Review of the Contents

I am the only one in the world. I'm a consulting detective.
—Sherlock Holmes, *A Study in Scarlet*

Early in the first adventure, Sherlock Holmes reveals his profession to his new roommate, Dr. John H. Watson. Eventually Watson describes sixty of Holmes's cases.[1] Murder is the most common offense, occurring in twenty-seven of the stories. Interestingly, the second most-common category is no crime at all. This happens in eleven stories.[2] The other twenty-two cases are scattered through thirteen other kinds of crime (Swift and Swift 1999).

The clients that consult Holmes come from a diverse set of backgrounds. They can be classified into eight types: business/professional (twenty-three), police (eight), damsel-in-distress (eight), landed gentry (eight), government (four), nobility (four), working class (three), none (two) (Swift and Swift 1999).

Of the thirty-seven times Holmes identifies the culprit, he decides to let him go free a surprising thirteen times. The other twenty-four are turned over to the police. A number of times the perpetrator dies before being caught. Interestingly Holmes claimed to have failed four times.[3] Obviously the reader can't know what to expect when even the masterful Holmes sometimes fails.

The use of so many different kinds of crime, so many types of clients, and so many different results, including failure, gives us a variety that keeps the stories fresh, even for rereading.

[1] Dr. Watson mentions numerous other Holmes cases in his narrations of the stories that comprise the Sherlock Holmes Canon. We deal only with the sixty stories published by Sir Arthur Conan Doyle.

[2] Others say that twelve stories involve no crimes (Berdan 2000).

[3] The number of failures depends on how one defines "failure" (Berdan 2000).

This work is about the science in the sixty Sherlock Holmes stories. Every story mentions something scientific. Many times it is just a molecule; sometimes it is a method. In some stories the science is of key importance. In others it just sets a mood. Those interested in science will nearly always find something of particular interest in a Sherlock Holmes story. Conan Doyle set out to write about a detective who actively employed science in his work. That he succeeded is not in dispute.

1

How Sherlock Holmes Got His Start

Section 1.1 Arthur Conan Doyle

Steel True, Blade Straight,

epitaph of Sir Arthur Conan Doyle

One can achieve somewhat of an understanding of how Sherlock Holmes came to exist by looking at the contributions of three people: Conan Doyle himself, Edgar Allan Poe, and Conan Doyle's mentor in medical school, Dr. Joseph Bell. First we shall look at Conan Doyle himself, focusing on those aspects of his life that led to his writing of the Sherlock Holmes stories.

Arthur Conan Doyle was born on May 22, 1859, in Edinburgh. His father, Charles Altamont Doyle, was English and his mother, Mary Foley, was Irish. His father had a drinking problem and was consequently less a factor in Conan Doyle's upbringing than was his mother. Charles would eventually end up in a lunatic asylum (Stashower 1999, 24). Mary Doyle instilled in her son a love of reading (Symons 1979, 37; Miller 2008, 25) that would later lead him to conceive of Sherlock Holmes. Conan Doyle's extensive reading had a great influence on the Sherlock Holmes stories (Edwards 1993). He was raised a Catholic and attended Jesuit schools at Hodder (1868–1870) and Stonyhurst (1870–1875), which he found to be quite harsh. Compassion and warmth were less favored than "the threat of corporal punishment and ritual humiliation" (Coren 1995, 15). Next he spent a year at Stella Matutina, a Jesuit college in Feldkirch, Austria (Miller 2008, 40). As Conan Doyle's alcoholic father had little income, wealthy uncles paid for this education. By the end of his Catholic schooling, he is said to have rejected Christianity (Stashower 1999, 49). At the less strict Feldkirch school, his drift away from religion turned toward reason and science (Booth 1997, 60). At this time he also read the writings of Edgar Allan Poe, including his detective stories. So, although Sherlockians debate the "birthplace" of Holmes, a claim can be made that Holmes was conceived in Austria.

In 1876, Conan Doyle began his medical studies at the highly respected University of Edinburgh. These years also played a large role in shaping the Holmes stories. One obvious factor was his continued exposure to science. Much of this book explores the presence of science in the sixty Holmes tales. The other significant factor from his medical studies was his mentor, Dr. Joseph Bell, whose deductions about patients impressed Conan Doyle to the extent that he added similar scenes in the Holmes tales. Upon completing his studies, Conan Doyle, now ready to set up a practice, headed to London for a meeting with his uncles. They could put him in a position to become a doctor to London's Catholic community through their many wealthy contacts. But he essentially threw that opportunity away by informing the family of his rejection of his Catholic upbringing. He was now, he told them, an agnostic, a term coined only a few years earlier by Thomas Huxley (Stashower 1999, 50). Conan Doyle knew what he was doing to his chances, but refused to pretend that he was still Catholic. As his epitaph suggests, his sense of honor would remain strong throughout his life. His uncles now refused to help him, and his career had a difficult beginning. Instead of London, Conan Doyle set up his medical practice in Southsea, Portsmouth, in 1882. In both his medical school thesis and other publications, Conan Doyle proved astute at understanding causes of diseases in ways not fully explained until much later (Miller 2008, 102). Although he continued to work there until 1890, he was not successful. His income the first year was £154, and it never rose much above £300 (Carr 1949, 66; Stashower 1999, 63). In fact, his first-year income tax return was sent back to him. The revenue inspector had written "Not satisfactory" on it. The quick-witted Conan Doyle resubmitted it unchanged with this notation: "I agree entirely" (Booth 1997, 96).

It was while in Portsmouth that Conan Doyle was first exposed to spiritualism. Although he would not publicly espouse it until 1917, eventually agnosticism would be discarded, and spiritualism would come to dominate his later life. Another important event during his Portsmouth years was his meeting Louisa Hawkins, known as "Touie." They met when he was called upon to give a second opinion of her brother Jack's diagnosis of cerebral meningitis. Conan Doyle took Jack Hawkins into his lodgings as a resident patient, but Jack died within a few days. The twenty-third Holmes tale would be titled *The Resident Patient* (RESI). Conan Doyle proceeded to court Touie, and they were married a few months later, on August 6, 1885. Because Touie had a small income of her own, Conan Doyle's poverty was somewhat relieved. But her health was very fragile, and she died at age forty-nine in 1906. Conan Doyle in the meantime had fallen in love with Jean Leckie, whom he had met in 1897. He is considered to have handled this delicate matter honorably. He married Jean fourteen months after Touie died (Stashower 1999).

Conan Doyle finally gave up the Portsmouth practice in 1890 when he went to Vienna for advanced study in ophthalmology. Upon his return he set up practice in London. He later wrote, "Not one single patient ever showed up." This gave rise to the well-known anecdote about him writing the Sherlock Holmes stories while waiting in his office for the patients who never came. As enticing as this story is, evidence exists that it might not be entirely accurate (Lellenberg et al. 2007, 291). Conan Doyle, a natural teller of tales, had already published several stories, beginning with *The Mystery of Sassassa Valley* in 1879. Now he decided to write a detective novel. Poe's detective, C. Auguste Dupin, would be his model. Holmes's intelligence would be so superior that he could solve mysteries that baffled others, but his solutions would be deduced. Chance, so common in the crime stories written between Poe's time (1841) and that of Conan Doyle (1887), would play no role. The result, *A Study in Scarlet* (STUD), was rejected by four or five publishers before Ward, Lock & Co. bought it outright for twenty-five pounds. It was published in *Beeton's Christmas Annual* for 1887. Conan Doyle never received any additional money from this story, which is still in print today. He later reported that STUD was not particularly well received in England, although it did go through several printings there.

But in America, Holmes was an immediate hit. STUD was well received in the United States. It actually "created an excited audience of Holmes fans" (Lachtman 1985, 14). So, conceived in Austria and born in London, Holmes was next resuscitated in America. Thus it was that in 1889, *Lippincott's Magazine*, published in Philadelphia, invited Conan Doyle and Oscar Wilde to meet in London (Coren 1995, 56). They shared a meal at the Langham Hotel with Lippincott's agent, Joseph Stoddart, and Irish MP Thomas Gill (Miller 2008, 119). Conan Doyle described the event as a "golden evening" (Green 1990, 1). The result was an agreement whereby each author would write a novel. Wilde proceeded to write his only novel, *The Picture of Dorian Gray*. Soon after the meeting, Conan Doyle submitted the name of his promised novel, *The Sign of the Six* (Booth 1997, 132). Conan Doyle had thought of his detective and decided to write the second Sherlock Holmes story. He even pays a bit of homage to Oscar Wilde by having one of the main characters, Thaddeus Sholto, resemble him. The title eventually became *The Sign of the Four* (SIGN). Like STUD it was one of the four long Sherlock Holmes stories. It has been argued that it was American interest that kept the Holmes saga going (Stashower 1999, 103).

With the third story, *A Scandal in Bohemia* (SCAN), Conan Doyle began his long series of Holmes short stories published in *The Strand Magazine*. It was the first of the fifty-six short stories, and it hit London like a bombshell. The circulation of magazine soared to 500,000 whenever a Holmes story was published (Riley and McAllister 1999, 24). The publisher, George Newnes, estimated that an extra 100,000 copies were sold whenever a Holmes tale appeared (Stashower 1999, 125; Miller 2008, 141). The small income of Dr. Conan Doyle now became a distant memory. But Conan Doyle tired of Sherlock Holmes quickly and considered killing him off in the eighth story. But Conan Doyle's mother was an ardent Holmes fan, and she commanded him not to do it. She even made a plot suggestion that he turned into *The Copper Beeches* (COPP), the fourteenth story (Stashower 1999, 126). But Holmes had to go. He was interfering with Conan Doyle's more serious literary efforts, namely his historical novels such as *Micah Clarke* (1889) and *The White Company* (1891). In addition, the task of devising new plots was becoming difficult. After borrowing from Poe in the first three stories, Conan Doyle repeats the same basic plot of keeping a young girl unmarried in order to retain control of her money in stories number five, *A Case of Identity* (IDEN), ten, *The Speckled Band* (SPEC), and fourteen, COPP. We get a fearsome stepfather in Dr. Grimesby Roylott in SPEC, a wimpy stepfather in James Windibank in IDEN, and a conniving father in Jephro Rucastle in COPP. The "feel" of each of these three stories is very different. The quality of the three is also extremely different. SPEC has been rated the best of the fifty-six short stories in every poll that has been taken. IDEN, with the same plot outline, has been described thusly: "The third story, IDEN, is a rather weak one" (Redmond 1981).

By the time he would finish, Conan Doyle would also repeat the theme of missing persons and have Holmes deal with six such cases (Lachtman 1985, 51–52). Additionally, in six stories—SIGN, *The Boscombe Valley Mystery* (BOSC), *The Five Orange Pips* (FIVE), "*The Gloria Scott*" (GLOR), *The Dancing Men* (DANC), and *Black Peter* (BLAC)—he reuses the idea of someone returning to England only to be followed and blackmailed or threatened (Schweickert 1980). So we find in December 1900, between writing stories number twenty-six, *The Final Problem* (FINA), and twenty-seven, *The Hound of the Baskervilles* (HOUN), that an article by Conan Doyle appears in *Tit-Bits* (Green 1983, 349). In it he says:

> When I had written twenty-six stories, each involving a *fresh plot*, I felt it was becoming irksome, this searching for plots.

That was one reason why in FINA, he has Holmes die in the clutches of archenemy Professor Moriarty as they both tumble over the Reichenbach Falls in Switzerland.

But when Conan Doyle brings Holmes back to life in the twenty-eighth story, *The Empty House* (EMPT), the problem of devising new plots continues. Utechin (2010, 32) has pointed out that the twenty-ninth, thirty-first, thirty-fifth, and fortieth stories all reprise themes he used in earlier Holmes tales, namely numbers three, nine, twenty-four, and twenty-five:

> *The Norwood Builder* owes much to *A Scandal in Bohemia*; *The Solitary Cyclist* has the plot of *The Greek Interpreter*; *The Six Napoleons* of *The Blue Carbuncle*; *The Adventure of the Second Stain* is a doublet of *The Naval Treaty*.

When Holmes "died" at Reichenbach Falls, the reaction in London was extreme. Black armbands of mourning were worn. Conan Doyle received numerous critical letters. Circulation of *The Strand Magazine* plummeted. Twenty thousand subscriptions were cancelled (Stashower 1999, 149; Miller 2008, 158). Ten years later, in 1903 in EMPT, we learn that Holmes had never fallen into the Reichenbach. Sherlockians refer to the ten-year period when Holmes was considered dead as the "Great Hiatus." Jean Leckie, later the second Mrs. Arthur Conan Doyle, had suggested the explanation for Holmes's escape from death (Booth 1997, 249). With the return of Holmes, the circulation of *The Strand Magazine* surged, and so did Conan Doyle's royalties. He could not afford to leave Holmes at the bottom of the Reichenbach; nor could he afford to remain a doctor. Never again did Conan Doyle allow Holmes to die. Holmes was still alive and tending bees in his retirement when, thirty-three stories later, Arthur Conan Doyle died on July 7, 1930. Along the way he had done more than create the greatest fictional detective ever. He had invented the literary device known as the "enigmatic clue" (Carr 1949, 350) with the famous Holmes quote from *Silver Blaze* (SILV), "The dog did nothing in the night-time." He had written the first "fool's errand" story,[1] *The Red Headed League* (REDH) (Priestman 1994, 315). And he had foreshadowed the "hardboiled detective" genre in *The Valley of Fear* (VALL) (Doyle and Crowder 2010, 183; Faye 2010, 15; Sullivan 1996, 170).

The path to the Sherlock Holmes stories then was this: a maternal influence toward voracious reading, strict Catholic schooling that drove him away from Catholicism, a love of science and reason acquired at school, the rejection by wealthy uncles because of his agnosticism, the failure of his medical practice, a natural talent for story-telling, Edgar Allan Poe's genius, Dr. Joseph Bell's brilliance, and the lucrative remuneration that kept Holmes alive.

Section 1.2 The Influence of Edgar Allan Poe

> *. . . his detective is the best in fiction.*
>
> —Sir Arthur Conan Doyle, *October 11, 1894, New York City*

Conan Doyle may have taken to writing as he waited in his office for the patients who rarely came, but Edgar Allan Poe "invented" the detective story when he published *Murders in the Rue Morgue* (RUEM) in 1841 (Silverman 1991, 171; Sova 2001, 66). At that time the word "detective" was not even in existence. Its

[1] In his continuing struggles to devise plots, Doyle uses the fool's errand theme in three stories, REDH, STOC, and 3GAR, the fourth, eighteenth, and fifty-third published.

first use was in 1843 (Silverman 1991, 173; Booth 1997, 104). In the forty years between Poe and Conan Doyle, there were many police stories, but they relied heavily on chance, guesswork, and deathbed confessions (Green 1987, 2). These stories "provided the bridge between Poe and the true tale of detection as created by Conan Doyle" (Cox 1993, xv). Then Conan Doyle, who clearly had read Poe, "reinvented" the detective story in 1887. In fact, initially there was a very heavy reliance on Poe. In the very first Holmes story, STUD, Conan Doyle borrows the concept of a cerebral detective with a sidekick sounding board. Thus arose the claim that Sherlock Holmes was modeled after Poe's C. Auguste Dupin, whose Watson counterpart is an unnamed narrator.

There were other influences on the first Sherlock Holmes story. The title is close to *L'Affair Lerouge*, Emile Gaboriau's 1866 story. The lengthy flashback is also found in Gaboriau (Edwards 1993, introduction to STUD, xxiv). Mormon killers are found here just as in Robert Louis Stevenson's *The Dynamiter* (Booth 1997, 104). Conan Doyle's concept of the American West in the second half of STUD drew on Mayne Reid's ideas (Edwards 1993, introduction to STUD, xxv). Even William Makepeace Thackeray is cited as a factor in shaping Conan Doyle's work (Edwards 1993, introduction to STUD, xv). But Poe was easily the biggest influence (Edwards 1993, STUD, xviii).

In the second Sherlock Holmes story, having made Holmes like a Poe character, Conan Doyle now reworks a Poe plot. The killer in Poe's RUEM is an Orang-Outang who scales an "unscaleable" wall, kills Madame L'Espanaye and her daughter, and then leaves by the same route. It was the first detective story (Silverman 1992, 174), as well as one of the earliest locked-room mysteries (Murphy 1999, 356). In SIGN, Conan Doyle writes his own locked-room story. He then has Tonga, a pygmy from the Andaman Islands, kill Sholto after matching the Orang-Outang's wall-scaling feat.

With the third Holmes tale, SCAN, Conan Doyle starts the hugely successful set of fifty-six Sherlock Holmes short stories. Again he reworks a Poe plot. In *The Purloined Letter* (PURL), a document belonging to a royal person is sought by the detective Dupin. The document is a compromising letter written by the queen of France. It is hidden in plain sight and recovered by the amateur sleuth using a ruse to divert attention so that he may take the letter and leave a substitute. The ruse is to distract minister D with a gunshot fired just outside his hotel room.

Holmes does likewise in SCAN, where the "document" is an incriminating photograph of the king of Bohemia and Irene Adler. The ruse is a cry of "fire!" plus a smoke bomb thrown in the window by Dr. Watson. Concerned about losing the photograph, Irene Adler's actions reveal to Holmes that the photograph is in her safe. In fact, Conan Doyle pokes fun at Poe, suggesting that a mere letter can never be as incriminating as a photograph:

King of Bohemia	There is the writing
Sherlock Holmes	Forgery
King	My private note paper
Holmes	Stolen
King	My own seal
Holmes	Imitated
King	My photograph
Holmes	Bought
King	We were both in the photograph
Holmes	Oh dear

There were similarities other than plot lines as well. Like Dupin, Holmes has eccentricities. Both authors used these eccentricities to make their characters more memorable. Once the Holmes tales became so popular, Conan Doyle had less need for eccentricities, and he had Dr. Watson wean Holmes from his drug habit. Dupin, though, remains unchanged, perhaps because, in only three stories, there was not enough time to have him evolve into something else. In addition, both sleuths are described as having "dual" natures. This is another instance where Conan Doyle borrowed from Poe. In Poe's tales, we read of Dupin's "bi-part soul." In Holmes, we see a man of intense action when on a case and the bored drug user whenever he misses the stimulation of work. "In his singular character the dual nature alternately asserted itself" (REDH). In the late 1800s when the Holmes stories were being published, the concept of the dual nature of humanity was the subject of much debate (Macintyre 1997, 222). The writings of Charles Darwin were relatively recent, and society was still digesting his ideas.

Conan Doyle also uses several literary devices found in Poe. One is the ruse just described. In addition to using such a ruse in SCAN, Conan Doyle does so again in *The Illustrious Client* (ILLU). In *The Norwood Builder* (NORW), a cry of "fire" along with an actual fire cause the culprit to leave the hiding place that Holmes has deduced is there. Another Poe idea is using newspapers to communicate with suspects by advertisements. In RUEM, Dupin advertises in *Le Monde* that an Orang-Outang has been found in the Bois de Boulogne. The sailor responds and is apprehended. Conan Doyle has Holmes advertise in newspapers beginning with the second story, SIGN. Sometimes he gets answers, such as Henry Baker responding to recover his Christmas goose in *The Blue Carbuncle* (BLUE). Other times there is no response, as in *The Naval Treaty* (NAVA). Even then, though, the absence of a response gives useful information to Holmes. All in all, newspapers are referred to in thirty-five of the sixty Holmes stories (Tracy 1977, 259).

Both Dupin and Holmes use disguises in their work. Twice in PURL Dupin dons green eyeglasses as a disguise, first in order to locate the queen's letter, and then to steal it. Again Conan Doyle immediately follows Poe and uses disguises in SIGN. When Mr. Windibank in IDEN wants to disguise himself so that his stepdaughter will not recognize him, he too chooses a pair of glasses, thick ones in this case. With a moustache and whiskers as well, Windibank is able to fool Mary Sutherland, even though she lives with him. Holmes uses disguises fourteen times in eleven different stories (Bunson 1994, 56). Conan Doyle may also have been influenced here by Emile Gaboriau's Monsieur Lecoq who, in *L'Affaire Lerouge* (1866), also uses disguises (Booth 1997, 106).

Another successful device that Holmes borrowed from Dupin was the habit of breaking in on Watson's train of thought. Dupin does just that in RUEM:

> Being both, apparently, occupied with thought, neither of us had spoken a syllable for fifteen minutes at least. All at once Dupin broke forth with these words:
>
> "He is a very little fellow, that's true, and would do better for the Theatre des Varietes."
>
> "There can be no doubt of that," I replied.
>
> "Dupin, this is beyond my comprehension. I do not hesitate to say that I am amazed. . . . "

There are several instances of Holmes reading Watson's mind. For example, in DANC:

> "So Watson, you do not propose to invest in South African Securities."
>
> "How on Earth do you know that?"
>
> "Now, Watson, confess yourself utterly taken aback."
>
> "I am."
>
> "I ought to make you sign a paper to that effect."
>
> "Why?"
>
> "Because in five minutes you will say that it is all so absurdly simple."
>
> "I am sure that I will say no such thing."

Upon hearing Holmes's explanation, Watson does declare the deduction to be absurdly simple.

Another example of Holmes breaking in on Watson's thoughts occurs in *The Adventure of the Cardboard Box* (CARD):

> "You are right Watson. It does seem a most preposterous way of settling a dispute."
>
> "Most preposterous."

Suddenly Watson realizes how Holmes had echoed the innermost thought of his soul:

> "What is this Holmes. This is beyond anything I could have imagined."

This time Watson confesses he is still amazed after Holmes explains how he traced Watson's thoughts.

One oddity of Poe's is the use of quotes from the classics at the openings of all four of his tales of ratiocination.[2] Conan Doyle adopts this approach in the early Holmes stories, using such quotes at the end. But after doing so in five of the first six stories, he returns to the practice only twice more, in stories written more than ten years later.

In both Conan Doyle and Poe, the official police force is not nearly as clever or as effective as the amateur. In fact, both amateurs criticize their predecessors: Dupin speaks ill of Vidocq; Holmes criticizes Dupin. Both authors have the relationship between the brilliant amateur and the official force undergo a similar evolution. In the first Dupin story, RUEM, he is resented by the prefect. In the second, *The Mystery of Marie Roget*, the prefect stops by to see Dupin, and in the third, PURL, the prefect actually gives the problem to Dupin. Initially there is hostility between Holmes and Scotland Yard. This is followed by cautious acceptance, full collaboration, and finally dependence (Dove 1997, 137).

Poe's influence on Conan Doyle was strongest in the early Holmes stories. But some of Conan Doyle's later Holmes tales also bear at least some resemblance to earlier Poe writings. In NAVA, Conan Doyle again returns to a missing document. As in Poe's PURL, the document could affect the government. Poe's *The Gold-Bug* (GBUG), though not a Dupin story, is often considered his fourth story of ratiocination. It appears to have influenced two of the Holmes tales, *The Musgrave Ritual* (MUSG) and DANC (Hodgson 1994, 213). These stories are discussed in the sections dealing with mathematics and cryptograms, respectively.

It should also be noted that some of Poe's non-detective writings appear to be an influence in the Holmes stories. In Poe's *Imp of the Perverse* (1845), an unnamed narrator commits murder using fumes from a poisoned candle. Conan Doyle's *The Adventure of Devil's Foot* (DEVI) involves two murders, by Mortimer Tregennis and of Mortimer Tregennis, with fumes from a root. *The Fall of the House of Usher* shares some elements with two separate Holmes stories. *Shoscombe Old Place* (SHOS) involves a brother with a dead sister and the fate of an estate (Fetherston 2006). *The Disappearance of Lady Frances Carfax* (LADY), like *Usher*, involves a case of premature burial (Vail 1996). So does MUSG, along with Poe's *The Premature Burial* and *The Cask of Amontillado* (CASK). Finally,

[2] A term used, often in referring to Poe's work, to describe tales wherein reasoning is a major factor.

there also seems to be some Poe influence in Conan Doyle's non-Holmesian work. In CASK, Fortunato is led into the wine cellar in Montresor's catacomb and sealed up by a wall, left there to die. In Conan Doyle's *The New Catacomb,* Kennedy is led into the newly discovered catacomb by Julius Burger. He is then left there to die, hopelessly lost in the pitch darkness of the cave, while Burger follows a string back to safety in the darkness. Some similarities between Poe's *The Gold-Bug* and Conan Doyle's first published story, *The Mystery of Sassassa Valley,* have been noted (Booth 1997, 62). Conan Doyle's *The Doings of Raffles Haw,* like Poe's *Von Kempelen and His Discovery,* deals with the "science" of alchemy (Stashower 1999, 117). Conan Doyle's Professor Challenger story, *The Poison Belt,* has been compared to Poe's *The Masque of the Red Death* (Redmond 1993, 79). And finally, in *The Horrors of the Heights,* Conan Doyle uses an airplane to travel to impossible elevations; a balloon does the same in Poe's *The Unparalleled Adventure of One Hans Pfaall.*

It is interesting to read how critics have responded to all of these Poe/Doyle comparisons. We conclude this section by looking at several comments on the two authors. Although all are in agreement that Poe was a large influence, we find some comments favorable to Conan Doyle and others not.

"Dupin is of little importance either in himself or in comparison to Poe, but Sherlock Holmes is greater than Conan Doyle" (Green 1987). Evidence for this lies in the fact that there is little or no interest in Dupin today, while Poe himself remains widely popular.[3] As Isaac Asimov points out, there are no societies devoted to the memory of Dupin, and few people remember Dupin, whereas Holmes is "a three dimensional living person" (Asimov 1987). A number of countries have expressed the same view by issuing stamps bearing the image and name of Sherlock Holmes, but ignoring Arthur Conan Doyle (Moss 2011). Most stamps depict Holmes in the famous deerstalker hat, which is more a creation of the artists who illustrated the stories than it is of Conan Doyle.[4]

> Perhaps the explanation for the immediate and lasting success is that Conan Doyle added humor and drama, both of which are lacking in Poe.
>
> It is impossible to read them (the three Dupin stories) without appreciating how much Conan Doyle improved upon the original formula. (Green 1987, 4)

[3] There is even a street named after him in New York City.

[4] Sidney Paget and later Frederic Dorr Steele, the two best-known illustrators, both depicted Holmes in a deerstalker hat.

> If you read Poe's three stories carefully you will find that the ingenious Dr. Doyle has picked him all to pieces, and worked up every available fragment with curious cleverness into his own stories. (Robert Blatchford, as cited in Green 1987, 9)
>
> ... used the same structure as Poe and virtually the same character, and that he copied, imitated, and plagiarized everything he felt was of value. The result was impressive. (Green 1987, 2)
>
> Conan Doyle was hardly able to string two or three words together or to use even the simplest idea without borrowing them. (Henri Mutrux 1977)
>
> *The Murders in the Rue Morgue* may be a classic locked room mystery, it may have the mind reading episode and one of the most memorable murders in detective fiction, but it is long-winded, intricate, and dull. (Green 1987, 4)

Asimov's opinion of Poe is that "he is passé, and much that he wrote, however admired by some, is simply unbearable to others" (Asimov 1987). Dorothy Sayers felt that Conan Doyle had improved on Poe's detective stories:

> He cut out the elaborate psychological introductions or restated them in crisp dialogue.
>
> He was sparkling, surprising, and short. (Sayers, ed. 1929)

An example of the long-winded Poe compared with the "crisp" Conan Doyle is found in Holmes's famous statement from *The Adventure of Beryl Coronet* (BERY), which is derived from this tedious statement in Poe's RUEM:

> Now, brought to this conclusion in so unequivocal a manner as we have been, it is not for us, as rational men, to reject it on account of apparent impossibilities. It is only for us to prove that these apparent impossibilities are, in reality, not such.

Holmes's succinct restatement in BERY:

> When you have eliminated the impossible, whatever remains, however improbable, must be the truth.

Several conclusions are warranted: First, Sherlock Holmes was based on Poe's Dupin. Second, although Poe is generally considered the greater author, Conan Doyle's detective fiction surpasses that of Poe. Third, Poe's non-detective writings are very highly regarded; Conan Doyle's are not.

Section 1.3 The Influence of Dr. Joseph Bell

Sherlock Holmes is the literary embodiment of a professor of medicine at Edinburgh University.

—Sir Arthur Conan Doyle, *May 1892*

Dr. Joseph Bell was born in Edinburgh in 1837 and spent his entire medical career in that city. Bell was known for his talents as a poet, naturist, and sportsman (Coren 1995, 22). He was a successful surgeon and editor of the *Edinburgh Medical Journal* for twenty-three years (Booth 1997, 49). Though never a faculty member at Edinburgh University Medical School, Bell did publish several textbooks. He also taught surgery at the Royal Infirmary. Conan Doyle, along with other med students, paid to attend his classes. Every Friday he held an outpatient clinic at the infirmary. There he would proceed to amaze both the students and the patients by his deductions. He was very successful in diagnosing the patient's conditions and sometimes their occupations, where they lived, and how they had traveled to the clinic. In 1878, Bell selected Conan Doyle to serve as his outpatient clerk for the Friday sessions (Booth 1997, 50). In this capacity, Conan Doyle became familiar with Bell's ability to observe trifles and make logical deductions from them.

One example involved a woman and her small child whom Bell had never met. After greeting one another, Bell displayed his deductions in a series of questions (Stashower 1999, 20).

"What sort of crossing did you have from Burntisland?"

"It was guid."

"And had you a good walk up Inverleith Row?"

"Yes."

"And what did you do with the other wain?"

"I left him with my sister in Leith."

"And would you still be working in the linoleum factory?"

"Yes, I am."

Bell had noted her accent, red clay on her shoes, a child's coat too large for the child with her, and dermatitis on the fingers of her right hand, a common condition in linoleum workers. Conan Doyle was impressed by this and other instances of Dr. Bell's brilliant deductions.

Another oft-quoted example of Dr. Bell in action deals with his instant diagnosis of a civilian patient's condition before even examining him.

> "Well, my man, you've served in the army."
>
> "Aye, Sir."
>
> "Not long discharged?"
>
> "Aye, Sir."
>
> "A Highland regiment?"
>
> "Aye, Sir."
>
> "A non-com officer?"
>
> "Aye, Sir."
>
> "Stationed at Barbados?"
>
> "Aye, Sir."

The observations that Dr. Bell used in this case were that, though the man was respectful, he did not remove his hat. They didn't remove hats in the army, but had he been long discharged, he would have adjusted to removing it. He had an air of authority, but not too strong an air; thus he was a non-commissioned officer. He was obviously Scottish and thus from a Highland regiment. And finally, his condition of elephantiasis was common in Barbados.

Conan Doyle had the Holmes brothers make similar deductions in *The Greek Interpreter* (GREE):

Mycroft	"Look at these two men who are coming towards us."
Sherlock	"The billiard-marker and the other?"
Mycroft	"Precisely. What do you make of the other?"
Sherlock	"An old soldier, I perceive."
Mycroft	"And very recently discharged."
Sherlock	"Served in India, I see."
Mycroft	"And a non-commissioned officer."
Sherlock	"Royal Artillery, I fancy."
Mycroft	"And a widower."
Sherlock	"But with a child."
Mycroft	"Children, my dear boy, children."
Watson	"Come, this is a little too much."

This, of course, is the scene that serves as one of the bases for the contention that, of the Holmes brothers, it was Mycroft who had the superior mind. Conan Doyle has Sherlock Holmes make brilliant deductions in several other stories. One famous example occurs in REDH when Holmes first meets his client, Jabez Wilson:

> Beyond the obvious facts that he has at some time done manual labor, that he takes snuff, that he a is a Freemason, that he has been in China, and that he has done a considerable amount of writing lately, I can deduce nothing else.

So it not surprising that Conan Doyle named Dr. Bell as the model for Sherlock Holmes. Conan Doyle first made this claim in an interview in May 1892. He said that Holmes was modeled after one of his teachers in medical school. In June 1892 in another interview, he named Bell as the model. When *The Adventures of Sherlock Holmes*, a book containing the first twelve short stories, was published in October 1892, Conan Doyle dedicated it to Dr. Bell (Green 1983, 17).

It has been noted that there was no mention of Bell in 1886 when Conan Doyle was beginning to create his detective. During these early days, as we have detailed in the previous section, Conan Doyle relied heavily on Poe. So Green concludes that Bell played a smaller role than Poe in the Holmes phenomenon (Green 1983, 28). Sir Henry Littlejohn was another of Conan Doyle's medical school instructors. In addition to lecturing at the medical school, he was police surgeon in Edinburgh. A forensic expert, he frequently served as an expert witness at trials. In fact, Dr. Bell served as an assistant to Dr. Littlejohn as official advisor to the British Crown in cases of medical jurisprudence (Liebow 1982, 119). Littlejohn is considered by some to have been as much a factor in the birth of the Holmes stories as was Bell (Jones 1994, 28). It is notable that, years after Bell's death in 1911, Conan Doyle himself mentioned Littlejohn as an important influence. In a speech in 1929, Conan Doyle named both Bell and Littlejohn as important in shaping his ideas (Green 1983, 27).

So, who was the model for Sherlock Holmes? Some say Conan Doyle himself was the real Holmes (Starrett 1930, 118). Certainly Conan Doyle's son Adrian believed his father was the real Sherlock Holmes (Liebow 1982, 224). In the 1940s, a public battle was waged in print over whether it was Dr. Bell or Dr. Conan Doyle who was Sherlock Holmes (Liebow 1982, 222–234). Dr. Bell's entertaining deductions show up in several of the Sherlock Holmes stories. But even these are foreshadowed by Poe in *The Man of the Crowd* (1840), where the unnamed narrator deduces occupations from the appearances of passers-by. The assertion that Holmes is a mixture of Poe's Dupin and Dr. Bell is undoubtedly correct (Booth 1997, 113). However, we feel that the few scenes based on Bell are hardly as influential as Poe's contributions: the very idea of a cerebral detective, the mind-reading episodes in Poe and Conan Doyle, the reworking of Poe plots from RUEM, PURL, and GBUG into SIGN, SCAN, and DANC. So, although Conan Doyle may have wanted to compliment his old mentors Bell and Littlejohn by naming them as models for Sherlock, it was Poe who influenced Conan Doyle most when he took up his pen to become a writer. Dr. Bell's important role was in giving Conan Doyle ideas about how to make his detective appear to be such a genius.

2

Meet the Main Characters

Section 2.1 Sherlock Holmes

He was the most perfect reasoning and observing machine that the world has seen.
—Dr. Watson, *A Scandal in Bohemia*

In this section we take a look at why Sherlock Holmes is one of the most recognizable characters in all of literature. Several factors contribute to this. After describing his physical characteristics and his personality, we look at the most important feature of his fame, his brilliant deductive abilities. It is in this that Arthur Conan Doyle is somewhat indebted to his mentor, Dr. Joseph Bell, as described in chapter 1.

In *A Study in Scarlet* (STUD), the very first Holmes tale, Dr. Watson describes Sherlock Holmes as being more than six feet tall, very lean, with piercing eyes and a thin hawk-like nose. Holmes's voice was high and occasionally strident. We learn later that his eyes were gray and he had a narrow face and black hair. Most illustrators over the years have faithfully reproduced this picture of the great detective.

Very little about Holmes's background is revealed to us. Most of what we do know is told in *The Greek Interpreter* (GREE). In this tale, the twenty-fourth of the sixty, Watson is shocked to learn that Holmes has a brother named Mycroft. It turns out that neither of the roommates has told the other that they have a brother. We also learn that the Holmes brothers are from a family of country squires. The family traces itself back to the Frenchman Horace Vernet (1789–1863), a noted painter of military scenes. Clearly there was enough money in Holmes's background for him to attend college. We know from *The "Gloria Scott"* (GLOR) that he did attend for two years.[1]

[1] Sherlockians have debated for over one hundred years whether Holmes attended Oxford or Cambridge, or perhaps even some other institution.

Figure 2.1 Sherlock Holmes, the world's first consulting detective

In *The Musgrave Ritual* (MUSG), Watson describes Holmes as very untidy. Apparently he kept his cigars in a coal scuttle and his tobacco in the toe of a Persian slipper. His correspondence was affixed to the mantel by a jackknife. In what is considered a patriotic gesture (Tracy 1977, 379), he honored his queen by using a pistol to shoot the letters VR, for Victoria Regina, into the wall of the Baker Street rooms. Though clearly not fussy about his chambers (see figure 2.1), Holmes is described in *The Hound of the Baskervilles* (HOUN) as committed to personal cleanliness.

Sherlock Holmes rarely exercised (*The Yellow Face,* YELL), but was still a good runner (HOUN), capable of a two-mile run when pursued (*Charles Augustus Milverton,* CHAS). An incident in *The Speckled Band* (SPEC) demonstrates Holmes's strength. The horrendous Dr. Grimesby Roylott bends Holmes's fireplace poker in an attempt to intimidate Holmes with a display of strength. After Roylott leaves, Holmes performs the even more difficult task of straightening the poker back to its normal shape. In *The Beryl Coronet* (BERY), Holmes claims, "I am exceptionally strong in the fingers." In several of the stories, we hear about Holmes the boxer. He tells Watson that he boxed in college (GLOR). Watson's opinion was that Holmes was an expert boxer (STUD and *The Final Problem,* FINA). In YELL, he calls Holmes "one of the finest boxers of his weight." In SIGN, we hear of Sherlock in action in the ring. McMurdo is the porter at Bartholomew Sholto's home and a prizefighter acquaintance of Holmes. Holmes greets him by saying, "Don't you remember that amateur who fought three rounds with you at Allison's rooms on the night of your benefit four years back." Several times his boxing talent was put to use in his detective work: He overcame a street "rough" in FINA and turned him over to police custody. He was twice able to "grass"[2] Joseph Harrison, who stole the treaty in *The Naval Treaty* (NAVA). Jack Woodley in *The Solitary Cyclist* (SOLI) had to be carted away after daring to fight with Holmes.

Conan Doyle was very interested in prizefighting. His successful novel *Rodney Stone* is said to have helped popularize boxing. In 1895, he was paid £4,000 in advance royalties, £1,500 for British serial rights, and £400 for American serial rights for *Rodney Stone*. The sum of £5,900 in 1895 was the equivalent of over £300,000 in 1995 (Booth 1997, 206). Conan Doyle's wealth was due to all of his writing efforts, not just to the Holmes stories.

Some have claimed that Holmes was a cold, hard person. This is based on several of Holmes's own statements. In *The Five Orange Pips* (FIVE), Holmes says, "I do not encourage visitors." In *The Devil's Foot* (DEVI), he states, "I have never loved." In SCAN we learn that he finds emotion to be abhorrent. In fact, in SIGN he says, "Love is an emotional thing, and whatever is emotional is opposed to that true cold reason which I place above all things." In *The Illustrious Client* (ILLU), he proudly proclaims, "I use my head, not my heart."

His personal traits, particularly the idiosyncrasies just described, make Holmes a memorable character. When Mrs. Hudson, his landlady, asks him when he would like to eat (*The Mazarin Stone*, MAZA), Holmes responds, "7:30, the day after tomorrow." He just can't be bothered with food when there is a culprit on the loose. His most pronounced trait was this "dual nature" (Tracy 1977, 163). It is mentioned in eight of the stories. Watson describes this in the very first tale, STUD:

> Nothing could exceed his energy when the working fit was upon him; but now and again a reaction would seize him, and for days on end he would lie upon the sofa in the sitting-room, hardly uttering a word or moving a muscle from morning to night.

An often-quoted example (Sweeney. in Putney et al. 1996, 43) of Holmes's duality comes from *The Red-headed League* (REDH). Watson contrasts Holmes the sleuth with Holmes the music lover:

> All the afternoon he sat in the stalls wrapped in the most perfect happiness, gently waving his long fingers in time to the music, while his

[2] Grass is an old sporting term meaning "to knock down." American editions read "grasp," which makes less sense since Harrison had a knife.

> gently smiling face and his languid dreamy eyes were as unlike those of Holmes the sleuth-hound, Holmes the relentless, keen-witted, ready-handed criminal agent, as it was possible to conceive.

This is another aspect of Poe's Dupin that Conan Doyle borrowed and inserted into his own creation. Dupin is described early in *The Murders in the Rue Morgue* (RUEM) as having a "bi-part soul."

Because the brilliance of Holmes is among the most important factors in the success of the stories, let us take a look at some examples that highlight this aspect of the character. In chapter 1, we saw several examples of how he was able to deduce Watson's train of thought, just as Dupin did in RUEM. Recall that when Young Stamford introduces Holmes and Watson in STUD, the very first words that Sherlock Holmes ever says to Dr. Watson are "How are you? You have been in Afghanistan, I perceive." Watson responds, "How on earth did you know that?" And we are off and running.

A Scandal in Bohemia (SCAN) is the first adventure following Watson's marriage to Mary Morstan, as described in the second story, *The Sign of the Four* (SIGN). No longer living with Holmes at 221B Baker Street, Watson stops by for a visit. Holmes remarks that he can tell that Watson has returned to practicing medicine, that he has been getting wet lately, and that he has a servant girl who is clumsy and careless. The accuracy of Holmes's deductions causes Watson to respond, "You would certainly have been burned had you lived a few centuries ago."

In *The Norwood Builder* (NORW), Holmes says to the stranger, John Hector McFarlane,

> You mentioned your name as if I should recognize it, but I assure you, that beyond the obvious facts that you are a bachelor, a solicitor, a Freemason, and an asthmatic, I know nothing whatever about you.

There are also numerous incidences of Holmes's ability to make amazing deductions from the most mundane of items. In HOUN, Holmes and Watson both try to deduce what they can from the walking stick that Dr. Mortimer had left at Baker Street the previous night. Neither Holmes nor Watson knows anything about Dr. Mortimer because they missed his visit.

Watson, using Holmes's methods to "read" the stick, concludes that Dr. Mortimer is a successful elderly man, well esteemed since the walking stick was a gift from the "members of C.C.H." The walking stick has been knocked about quite a bit; for example, the iron ferrule is worn. Watson also concludes that Dr. Mortimer is a country practitioner who does a lot of walking. The C.C.H. engraved on a silver band refers to the local hunt club, whose members have given Dr. Mortimer the stick in appreciation for his medical work.

Figure 2.2 Holmes uses his lens to examine Dr. Mortimer's walking stick in *The Hound of the Baskervilles.*

Holmes's analysis is somewhat different (see figure 2.2). "I am afraid, my dear Watson, that most of your conclusions were erroneous." Holmes agrees that Dr. Mortimer is a country practitioner who does a lot of walking. But C.C.H. stands for Charing Cross Hospital in London. The stick was a gift upon Dr. Mortimer's leaving London to practice in the country. Reasoning that most doctors would not give up a position at Charing Cross Hospital for one in the countryside, Holmes deduces that Mortimer actually held a lowly position in London and was probably little more than a student there. Thus he expects to meet a young doctor, not Watson's predicted elderly man. Holmes also claims that Mortimer owns a middle-sized dog. This last observation makes Watson laugh. Naturally, when Dr. Mortimer returns, we find that Holmes was right. He was able to make that deduction from the dog's teeth marks on the walking stick.

Holmes has another opportunity to demonstrate his deductive powers in YELL. Once again a potential client leaves an item at the Baker Street lodgings. This time it is a pipe. Holmes spares Watson the embarrassment of being out-deduced and proceeds to directly interpret Grant Munro's pipe. Holmes concludes that Munro is a muscular man, left handed, careless, well to do, with an excellent set of teeth, who highly values the pipe. The bases for these conclusions are that he was strong enough to bite through the amber pipe stem and careless in that he managed to char the pipe by holding it too near to a gas jet to light the expensive tobacco within. That Munro highly values the pipe is clear by the fact that he had it repaired twice, both times at a cost nearly equal to the purchase price of a new pipe.[3]

In *The Golden Pince-Nez* (GOLD), Holmes does a brilliant analysis of the pince-nez spectacles found clutched in the dead man's hand. Stanley Hopkins, a Scotland Yard detective who appears in stories thirty-three, thirty-seven, thirty-eight, and thirty-nine, shows the glasses to Holmes. Holmes presents the amazed Hopkins with a handwritten note containing a detailed description of the owner.

> Wanted, a woman of good address, attired like a lady. She has a remarkably thick nose, with eyes which are set close upon either side of it. She has a puckered forehead, a peering expression, and probably rounded shoulders. There are indications that she has had recourse to an optician at least twice during the last few months. As her glasses are of remarkable strength, and as opticians are not very numerous, there should be no difficulty in tracing her.

This so astonishes Hopkins and Watson that Holmes provides an explanation to them. Such delicate and expensive glasses would only belong to a well-to-do woman. Holmes deduced from repairs made to the cork linings on the clips that she had visited an optician twice recently. The width of the clips meant a broad nose. The position of the lenses indicated that her eyes were set close to her nose. Holmes associated a puckered forehead, peering expression, and rounded shoulders with the need for such strong glasses.[4]

The Blue Carbuncle (BLUE) has a scene that shows Holmes at his deductive best. Petersen, a commissionaire,[5] has retrieved a Christmas goose and a battered felt hat following an incident in the early morning hours of Christmas day. The goose bears a tag that reads "For Mrs. Henry Baker." The hat has the initials H. B. written on the inside. Holmes gives Watson a chance to interpret the hat, and Watson responds: "I can see nothing." After pointing out that Watson sees exactly what he sees, Holmes proceeds with his analysis of Henry Baker's hat.

Holmes concludes that the owner is highly intellectual, was formerly well-to-do, but is no longer. He used to have foresight, but displays a moral retrogression that is probably due to alcohol. His wife has ceased to love him. He has retained some

[3] Holmes makes an interesting statement about the amber stem. His words differ in the American and English editions of YELL. This is discussed in the appendix (see "Doyle Scams").

[4] If this sounds farfetched, remember that Conan Doyle was an ophthalmologist.

[5] A uniformed military veteran employed in a variety of tasks.

self-respect, and is sedentary and middle-aged. His hair was recently cut and had lime cream applied to it. Holmes's final flourish, "It is unlikely that he has gas laid on in his house," causes Watson's response, "You are certainly joking."

The large size of the hat leads Holmes to say that Baker was intellectual. Here he is claiming that a large head means a large brain and thus increased mental capacity. Holmes sort of refutes his own idea when he tries on Baker's hat and it "came right over the forehead and settled upon the bridge of his nose." Surely Sherlock Holmes has more mental capability than Henry Baker. The fact that the concept of a "brain attic" that can get filled up was endorsed by Oliver Wendell Holmes may have been where Conan Doyle got the idea (Moss 1991). This idea also harkens back to Holmes's comment in STUD that one's brain can get filled up. It is a size issue, and he does not wish to clutter his mind with useless facts. So when Watson informs him that the earth revolves around the sun, Holmes declares he will do his best to forget that useless fact! Because Baker's hat is an expensive one, albeit out of date, Holmes deduces that he formerly had money, but no longer does. Foresight is evident from the fact that Baker bought a hat securer to protect his hat from the wind. The moral retrogression is suspected because the securer is broken and has not been replaced.

In a piece of reasoning that would now be seen as sexist, Holmes concludes that Mrs. Baker has ceased to love her husband by the fact that she has not brushed his very dusty hat. The predictions of a sedentary middle-aged man with a recent haircut and lime cream come from stains on the lining of the hat. That Henry Baker has retained some measure of self-respect is clear to Holmes from the fact that Baker has tried to conceal those stains. It is the presence of five tallow stains that leads Holmes to conclude that Baker has no gas "laid on in his house." Holmes certainly wasn't joking, and it all proved to be correct.

The famous Sherlockian scholar Christopher Morley had a strong opinion about the relative merits of Christmas stories by Conan Doyle and Dickens: "I am quite serious when I say that, as a story, *The Blue Carbuncle* is a far better work of art than the immortal *Christmas Carol*" (Rothman 1990, 118).

As a final example of Holmes reading objects, consider the incident in SIGN, the second story, where Watson decides to test the deductive ability of his relatively new roommate. Watson's challenge: "I have a watch here which has recently come into my possession. Would you have the kindness to let me have an opinion upon the character or habits of the late owner?" Holmes examines the watch and responds:

> I should judge that the watch belonged to your elder brother.
>
> He was a man of untidy habits—very untidy and careless. He was left with good prospects, but he threw away his chances, lived for sometime in

> poverty, with occasional short intervals of prosperity, and finally, taking to drink, he died.

Watson reacts strongly:

> This is unworthy of you Holmes. I could not have believed that you would have descended to this. You have made inquiries into the history of my unhappy brother, and now you pretend to deduce this knowledge in some fanciful way. You cannot expect me to believe you have read all this from his old watch!

The scene closes with a sentence that all science teachers must love:

> "My dear doctor, pray accept my apology. I assure you that I did not even know that you had a brother until you handed me the watch."
>
> "But it was not mere guesswork?"
>
> "No, no: I never guess. It is a shocking habit, destructive to the logical faculty."

As he did with objects, Holmes could also make deductions about people. We get our first look at this in the beginning of the first adventure, STUD. Inspector Lestrade has found the letters "RACHE" on the wall in blood (see figure 2.3). He is convinced that a woman named Rachel is the key to solving the case. Holmes, though, spends twenty minutes examining the room from every angle using a magnifying glass and a tape measure. As he is about to leave, he informs the Scotland Yard inspectors Lestrade and Gregson that

> [t]he murderer was a man. He was more than six feet high, was in the prime of life, had small feet for his height, wore coarse, square-toed boots and smoked a Trichinopoly cigar. He came here with his victim in a four-wheeled cab, which was drawn by a horse with three old shoes and a new one on his off foreleg. In all probability the murderer had a florid face, and the fingernails of the right hand were remarkably long.

Holmes concludes:

> Rache is the German word for revenge; so don't lose your time looking for Miss Rachel.

At the beginning of SCAN, the third story, Holmes receives a curious note:

> There will call upon you tonight, at a quarter to eight o'clock, a gentleman who desires to consult you upon a matter of the very deepest moment. Your recent services to one of the royal houses of Europe have

Figure 2.3 Inspector Lestrade triumphantly points out his discovery to Sherlock Holmes.

> shown that you are one who may be trusted with matters which are of importance which can hardly be exaggerated. This account of you we have from all quarters received. Be in your chamber then at that hour, and do not take it amiss if your visitor wear a mask.

When Watson asks Holmes what he thinks of the note, Holmes's answer shows his reliance on the scientific method:

> It is a capital mistake to theorize before one has data. Insensibly one begins to twist facts to suit theories, instead of theories to suit facts.

Another instance of Holmes's commitment to the scientific method occurs in *The Sussex Vampire* (SUSS) when he says, "One forms provisional theories and then waits for time of fuller knowledge to explode them."

At the beginning of SCAN, Holmes proceeds to analyze the stationary on which the king had written the note. It is obviously expensive. Holmes decides the writer must be a German because of the sentence construction. Reading "This account of you we have from all quarters received," Holmes declares, it is only "the German who is so uncourteous to his verbs." When he finds the letters "Eg," "P," and "Gt" woven into the texture of the notepaper, Holmes deduces that

Gt stands for Gesellschaft, signifying a company in German, and P is for Papier. Then by consulting his *Continental Gazetteer*, he finds that the Eg indicates Egria, a part of Bohemia. Hearing horses on Baker Street, Holmes looks out the window and notes his masked visitor arriving in a very expensive rig with horses. "There's money in this case, Watson, if there is nothing else."

Finally, in the fourth tale, REDH, Holmes is again confronted with curious facts at the start. Why was Jabez Wilson hired to spend the hours from 10 AM to 2 PM each day away from his pawn shop copying every word in the *Encyclopedia Britannica*? And four pounds per week seems a generous salary for such menial work. And why was it that he only got the job because of his fine red hair. When Wilson arrives at his copying job one Friday, he is distressed to find a note: "The Red-Headed League is dissolved." Reluctant to give up the easy income, Wilson consults Holmes.

Holmes is delighted by the unusual circumstances. He tells Wilson, "I really wouldn't miss your case for the world." Later he says to Watson, "It is quite a three pipe problem." Holmes has learned from Wilson that his assistant, Vincent Spaulding, spends as much time as he can in the basement of the pawn shop. Reasoning that the purpose of the "fool's errand" is to allow Spaulding more freedom in the basement, Holmes deduces that a tunnel is being dug. He visits the premises and sharply strikes the pavement in front of the shop with his stick. The sound tells him that the tunnel is headed in another direction. When he notices the City and Suburban Bank nearby, Holmes realizes what is up. To verify his conclusion, he knocks on the door and Spaulding answers. While asking for directions, Holmes peers at the knees of Spaulding's trousers. Seeing the signs of dirt that he expected, Holmes and the police are waiting that Saturday evening as Vincent Spaulding, a.k.a. John Clay, the fourth-smartest man in London, tunnels into the bank's vault.

It has been over 125 years since Sherlock Holmes appeared. The stories have never gone out of print. Of all the reasons for this, it is the vivid characterization of Holmes that is most important. His brilliant deductions continue to amaze and amuse today's readers. Did Holmes use these amazing deductive powers to solve crimes? Of course he did, and his forensic methods are discussed at length in chapter 3.

Section 2.2 Dr. John H. Watson

Watson is Conan Doyle's great creation.

—C. R. Putney et al., *Sherlock Holmes: Victorian Sleuth to Modern Hero*

Sherlock Holmes and Dr. Watson rank as one of the greatest duos in literature (see figure 2.4). They are mentioned along with such famous pairs as the Lone Ranger

Figure 2.4 Holmes and Watson, the most famous duo in literature

and Tonto and Han Solo and Chewbacca (Skene-Melvin in Putney et al. 1996, 122). They surpass other detective pairs such as Nero Wolfe and Archie Goodwin, Nick and Nora Charles, and Charlie Chan and his number-one son. We trace Dr. Watson's background to see how the partnership came to be.

Watson attended school in England, receiving a degree in medicine from the University of London in 1878. He then worked as a staff surgeon at St. Bartholomew's Hospital (usually referred to as Bart's). Next he joined the army medical department and took additional training as a military surgeon. Britain was then involved in what was known as the Second Afghan War (1878–1880) (Klinger 2006, 10). Watson was attached to the Fifth Northumberland Fusiliers. The battle of Maiwand on July 27, 1880, was a decisive defeat for the greatly outnumbered English forces (www.britishbattles.com). At Maiwand Watson was struck by a jezail[6] bullet, and his life was saved by Murray, his orderly. In the first Holmes story, STUD, Conan Doyle places the wound in the shoulder; in the second, SIGN, he puts it in the leg.[7] Upon recovering from the wound, Watson contracted "enteric fever." Now in

[6] A jezail is a heavy, long-barreled musket.

[7] Delighted Holmesians still argue about the position of Watson's wound.

very poor health, he was sent back to England to recover. At this time, he is described as extremely thin and well tanned.

The government funded his recovery with a wound pension of a mere eleven shillings and sixpence a day. He gravitated to London despite having no family in England (suggesting Scottish ancestry). Unemployed, he soon found his financial situation difficult. He decided that he could no longer afford to reside at a private hotel on the Strand. The very day he came to this conclusion, he chanced on "Young Stamford" at the Criterion Bar. Stamford had been his dresser[8] at Bart's. When Watson mentioned that he was looking for cheaper lodgings, Stamford told him about another person who was doing the same. He then took Watson to Bart's where Holmes was doing some research. Sherlock's first words to Watson are a deduction: "How are you? You have been in Afghanistan, I perceive."

So Stamford plays a vital role in the first story, STUD, even though we never hear of him again in the other fifty-nine tales. Before parting company with Stamford, Watson asks him, "[H]ow the deuce did he know I had come from Afghanistan?" With a smile, Stamford challenges Watson to "study him." But he predicts, "I'll wager he learns more about you than you about him."

Not long after moving into their lodgings at 221B Baker Street, Watson gets involved in Holmes's cases. In fact, he marries Mary Morstan, Holmes's client in the second story, SIGN. Eventually he becomes the chronicler of fifty-six of the sixty stories.[9] Holmes doesn't always appreciate his writing efforts. In *The Abbey Grange* (ABBE), Holmes complains,

> Your fatal habit of looking at everything from the point of view of a story instead of as a scientific exercise has ruined what might have been an instructive and even classical series of demonstrations. You slur over work of the utmost finesse and delicacy, in order to dwell upon sensational details which may excite, but cannot possibly instruct, the reader.

The irritated Watson fires back, "Why do you not write them yourself." Again in *The Copper Beeches* (COPP), Holmes says, "You have degraded what should have been a course of lectures into a series of tales." When Holmes does serve as narrator, in *The Blanched Soldier* (BLAN), he learns a good lesson: "I am compelled to admit that, having taken my pen in hand,

[8] Surgeon's assistant.

[9] The forty-eighth and forty-ninth stories, LAST and MAZA, are written in the third person. Holmes narrates the fifty-sixth and fifty-seventh stories, BLAN and LION.

I do begin to realize that the matter must be presented in such a way as may interest the reader."

Though we learn in the first story, STUD, that Watson is thin and tan, it isn't long before his appearance changes. In *The Boscombe Valley Mystery* (BOSC), the sixth story, Mrs. Watson remarks, "You have been looking a little pale lately." He has also regained his weight. In CHAS, he is described as being of middle size, strongly built, with a square jaw, thick neck, and a moustache. Although somewhat athletic, in SIGN, Watson is limping—Conan Doyle has moved the Afghan wound to his leg. But by the time of HOUN, Watson tells us that he is fleet of foot. Indeed, at the end of CHAS, he and Holmes run for two miles after leaving Milverton's house. Of course, they are fleeing the police and are thus motivated to keep running!

In *Shoscombe Old Place* (SHOS), the very last of the sixty Holmes stories to be published (in 1927), we learn that Watson's wound pension is still being paid. He admits that he wagers about half of it on horse races. In DANC, we learn that Watson's checkbook is locked in Holmes's drawer. Some have speculated that Watson's betting on the ponies was out of control at this point. So a picture of a sporting man emerges. In SUSS, Watson recognizes the name of former rugby star Bob Ferguson, "the finest three-quarter Richmond ever had." Watson himself had played rugby for Blackheath, the premier rugby club of England (Tracy 1977, 37).

Apparently Watson is a handsome man. He boasts of "experience of women which extends over many nations and three separate continents" (SIGN). In *The Second Stain* (SECO) Holmes tells Watson, "[T]he fair sex is your department." And in *The Retired Colourman* (RETI), Holmes refers to Watson's "natural advantages" with women. The manners in which Holmes and Watson describe women clearly show the difference between the two men. At the beginning of SIGN, Holmes and Watson consult with Mary Morstan, who later becomes Mrs. Watson. This is how they respond to her: "What a very attractive woman," says Watson. Holmes responds, "Is she? I did not observe." But when it comes to detecting, their powers of observation are reversed. Holmes, who never misses a clue, can be very critical of Watson: "You see, but you do not observe, the distinction is clear."

Watson is always very interested in the female shape (Nightwork a.k.a. Morley, in Shreffler 1989, 190). Here is how he describes some of the women in the tales:

Irene Adler (STUD): Her superb figure outlined against the lights
Mrs. Neville St. Clair (TWIS): her figure outlined against the flood of light
Grace Dunbar (THOR): a brunette, tall, with a noble figure
Isadora Klein (3GAB): a perfect figure
Mrs. Merrilow (VEIL): buxom landlady type
Eugenia Ronder (VEIL): full and voluptuous

Lady Brackenstall (ABBE): I have seldom seen so graceful a figure

Lady Hilda Trelawney Hope (SECO): the most lovely woman in London

[and] a queenly figure

Holmes on Lady Hilda: Think of her appearance Watson—her manner, her suppressed excitement, her restlessness, her tenacity in asking questions

For Holmes, it is a question of a woman's appearance giving a clue; for Watson, what matters is what kind of body she has.

Much has been written about Watson's skill as a physician. It turns out that Dr. Watson often administers to people while participating in Holmes's cases. A number of times, Watson is called on to revive someone in need. In the case of Miss Barnet in WIST, he merely used strong coffee to rouse her from opium poisoning. More often it was brandy that was administered. James Ryder in BLUE, Thornycroft Huxtable in *The Priory School* (PRIO), Victor Hatherly in *The Engineer's Thumb* (ENGR), and Mr. Melas in GREE were all given spirits to revive them. The most dramatic use of brandy occurs in NAVA when Holmes shocks Percy Phelps with the return of the missing treaty. Lacking today's medications, brandy was a common and reasonable choice. It served as "a restorative, as a tranquilizer, as a pain reliever" and "as a means of reviving" (Scholten 1988).

There are several other instances of Dr. Watson in action. He dressed the thumb of Victor Hatherly in ENGR. He twice administered what has been described (Simpson 1934, 55) as artificial respiration: to the crook Beddington in STOC and to Lady Frances Carfax in *The Disappearance of Lady Frances Carfax* (LADY). When Kitty Winter threw sulphuric acid in the face of Baron Gruner (ILLU), Watson did what he could for the Baron, including giving him an injection of morphia.

This medical work indicates that Watson was a competent generalist. That he was more can be seen by noting that he made efforts to keep current in medicine. He is known to read the *British Medical Journal* (*The Stockbroker's Clerk*, STOC), still today a highly respected source of medical information. In SIGN, we find him reading up on "the latest treatise on pathology.", He reads about surgery in GOLD, nervous lesions in *The Resident Patient* (RESI), tropical disease in *The Dying Detective* (DYIN), and French psychology in *The Six Napoleons* (SIXN).

The well-read Watson was sometimes able to make reasonable diagnoses from visual observation alone. In SUSS, he can see that young Jacky Ferguson had a "weak spine." In SPEC, he detects that Dr. Grimesby Roylott had what has been termed a "bilious condition" (Simpson 1934, 48). Watson could tell at a glance that Isa Whitney was a drug addict (*The Man With the Twisted Lip*, TWIS). His diagnosis of aortic aneurism for Jefferson Hope in STUD has been criticized.

But Thaddeus Sholto's anxiety in SIGN did not fool him. A brief examination allowed Watson to inform Sholto that there was nothing wrong with his heart. He was perhaps stepping outside his area of expertise when he made a diagnosis of "monomania" in SIXN, and it was based on his misinterpretation of the clues in the case. But there is at least one opinion that Watson was also capable in the area of mental problems (Kellogg 1989).

Some have said his knowledge of first aid shows him "at his best" (Simpson 1934, 54). Others have said his knowledge of first aid was "nil" (Suszynski 1988, 15). Perhaps his greatest medical achievement was weaning Holmes from his drug habit. It has been pointed out that it took Watson eight years to achieve (Suszynski 1988, 13). But altering the behavior of such a forceful personality as Sherlock Holmes would always be a formidable challenge. In the opening scene of SIGN, Watson asks with disgust, "Which is it today, morphine or cocaine?" It was the famous 7 percent solution of cocaine.[10] In the first Holmes story, STUD, Watson suspects that Holmes is "addicted to the use of some narcotic." This is immediately confirmed in the opening paragraph of the second story, SIGN:

> Sherlock Holmes took his bottle from the corner of the mantel-piece, and his hypodermic syringe from its neat morocco case. With his long, white, nervous fingers he adjusted the delicate needle and rolled back his left shirtcuff. For some little time his eyes rested thoughtfully upon the sinewy forearm and wrist, all dotted and scarred with innumerable puncture-marks. Finally he thrust the sharp point home, pressed down the tiny piston, and sank back into the velvet-lined armchair with a long sigh of satisfaction.

Watson had watched this ritual three times a day for many months. Cocaine had debuted as a "wonder anaesthetic" in 1884, only three years before Holmes and Watson first met (Smith 2011, 69). Sigmund Freud began treating patients with cocaine that same year (Riley and McAllister 1999, 88). Both morphine and cocaine were perfectly legal at the time (Doyle and Crowder 2010, 45). However, Watson was among those who, early on, saw the danger in cocaine usage. Watson determines to wean Holmes from the habit. In *The Missing Three-Quarter* (MISS), we learn that he succeeded.

We've looked at Watson's background, appearance, and medical skills. We conclude by examining the feature that makes him the beloved character he has become: his loyal service to Holmes (see figure 2.5). One obvious aspect of his devotion to Holmes is his willingness to put his own interests aside and

[10] Its fame is due mainly to Nicholas Meyer's 1974 book, The Seven-Per-Cent Solution.

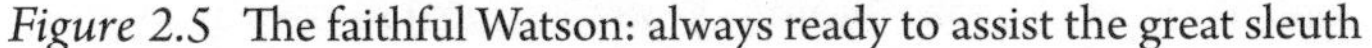

Figure 2.5 The faithful Watson: always ready to assist the great sleuth

do Holmes's bidding, no matter what. In *The Creeping Man* (CREE), Holmes implores Watson,

> Come at once if convenient—if inconvenient come all the same.

In ABBE,

> Come, Watson, come.... The game is afoot. Not a word! Into your clothes and come!

And Watson always responds favorably: "Count me in, Holmes" (MAZA).

In at least five instances, Watson agrees to go alone to work on one of Holmes's cases. In HOUN, it is initially Watson who goes to Baskerville Hall to investigate the death of Sir Charles Baskerville. In SOLI, Holmes sends Watson to check on Violet Smith's story. When Lady Frances Carfax disappears, Watson willingly heads off to Switzerland to seek her in Lausanne. In another instance (ILLU), Watson studies Chinese pottery for a week. Then, posing as an expert, he visits Baron Gruner, a collector, in order to distract him. Gruner recognizes that Watson is a fraud. But no matter, the baron is occupied long enough for Holmes to steal his diary. The goal is to prevent the marriage of the unsuspecting Violet de Merville to the evil baron. In RETI, the fifty-eighth story published, Holmes is still sending Watson on missions. This time, upon hearing Watson's report on his trip to visit Josiah Amberly in Lewisham, Holmes remarks, "It is true that you have missed everything of importance." In LADY, Holmes says, "I cannot at the moment recall any possible blunder which you have omitted." So the good doctor is always willing to help, though not always able or appreciated.

In a number of other cases, Watson willingly accompanies Holmes to a great variety of places. In PRIO, he goes with Holmes to the north of England. In 3STU, he spends some weeks with Holmes in a "university town." In GOLD, Watson takes the train to Chatham. In FINA, he accompanies Holmes to the continent as they flee from Professor Moriarty. In SHOS, he goes with Holmes to the Green Dragon Inn in Berkshire. At the end of BLAC, he even heads off with Holmes to Norway for a few weeks. When Holmes sends a telegram in BOSC, Watson's wife urges him to go join Holmes in Herfordshire. In CHAS, he is so determined to be the faithful helper that he threatens to inform the police of Holmes's plans to burgle Milverton's house—unless he too can be in on the burglary. No more faithful helper can be imagined.

The Milverton burglary is just one example where Watson is willing to face danger on Holmes's behalf. In SIGN, the poison darts of Tonga, the Andaman Islander, endanger both Holmes and Watson. In SPEC, there are cheetahs and baboons roaming the grounds at Stoke Moran. Watson pulls Holmes away from the fumes that threaten both their lives in DEVI. There are several cases where he carries his revolver for fear that the situation may turn dangerous. Consider Holmes's note sent to Watson in *The Bruce-Partington Plans* (BRUC):

> Am dining at Goldini's Restaurant, Gloucester Road, Kensington. Please come at once and join me there. Bring with you a jimmy, a dark lantern, a chisel, and a revolver.

In some cases, such as REDH, SPEC, and *The Problem of Thor Bridge* (THOR), Watson doesn't use his gun. In COPP, he shoots and kills Carlo, the mastiff that has his owner, Jephro Rucastle, by the throat (see figure 2.6). In SIGN, he fires at Tonga, the pygmy/murderer. In BLAC, the revolver is held to the temple of Patrick Cairns. In *The Empty House* (EMPT), Watson hits Colonel Sebastian Moran, Professor Moriarty's chief of staff and the "second most dangerous man in London," with the butt of the revolver.

The opening quote for this chapter is clearly wrong. It is Sherlock Holmes who is Conan Doyle's great creation. Time after time, Holmes has been voted fiction's greatest detective. Holmes, not Watson, revived the genre after forty years of languishing following Poe's stories. Initially Watson exists in the Sherlock Holmes tales to fill the role played by the unnamed narrator in Poe's three Dupin stories. But, just as Holmes surpasses Dupin, Watson is also a more vivid character than his counterpart in Poe's work. He is presented as

Figure 2.6 Dr. Watson kills Carlo the mastiff in *The Copper Beeches.*

loyal helper, friend, and chronicler. If Conan Doyle intended Watson to be Holmes's "rather stupid friend" (Smith 2011, 33), then here is an instance where the author failed. As we've seen, the result is a complex character of substance. Though Holmes dominates, the Canon would be a lesser work if there were no Dr. Watson.

Section 2.3 Professor James Moriarty

He is the Napoleon of crime, Watson.

—Sherlock Holmes, *The Final Problem*

Sherlock Holmes's greatest enemy was Professor James Moriarty (see figure 2.7), who has been called the first great fictional master criminal (Smith 2011, 122), one of the most memorable antiheroes in all of literature (Macintyre 1997, 222) and the greatest villain in all of detective literature (Doyle and Crowder 2010, 128). Conan Doyle manages to provide the professor with an air of malevolence. Part of that is due to his eerie appearance. Moriarty is described as tall, thin, and pale, with sunken grey eyes and a domed forehead. His face protrudes and oscillates in a reptilian fashion.

After years as a consulting detective, Holmes has sensed a central force dominating the London crime scene. "He is the Napoleon of crime," he tells Watson (FINA). He has finally decided it is Moriarty at the center of a large crime

Figure 2.7 Professor Moriarty, Holmes's greatest adversary

organization. Holmes has become so familiar with Moriarty's devious ways that he can recognize his crimes.

> You can tell an old master by the sweep of his brush.
> I can tell a Moriarty when I see one. (*The Valley of Fear*, VALL)

Before Holmes meets the professor, he has visited Moriarty's rooms three times (VALL). But he finds nothing incriminating. Continuing his efforts, Holmes is finally ready, in *The Final Problem* (FINA), to spring a trap on the Moriarty organization.

The professor's dominant feature is his great mental capacity. His brilliance is such that Sherlock Holmes admits to Watson that Moriarty is his intellectual equal. In his first appearance, in FINA, Moriarty has the audacity to show up at Holmes's lodgings at 221B Baker Street. He has sensed Holmes's trap. The vivid scene that results shows the two great minds dueling:

> HOLMES: I can spare you five minutes if you have anything to say.
> MORIARTY: All I have to say has already crossed your mind.

> HOLMES: Then possibly my answer has crossed yours.
> MORIARTY: You stand fast?
> HOLMES: Absolutely.

That Holmes considers Moriarty his greatest challenge is made evident at the end of their interview. Moriarty warns that if Holmes brings him down, he will in turn destroy Holmes. Sherlock's response:

> [I]f I were assured of the former eventuality, I would cheerfully accept the latter.

Moriarty's name comes up in only seven of the Holmes stories. In four of those, NORW, MISS, *His Last Bow* (LAST), and ILLU, it is a mere mention. In FINA, after confronting Holmes in his Baker Street lodgings, Moriarty chases Holmes to Switzerland where he locates him at the Englisher Hof in Meiringen. They then have their famous struggle at the top of the Reichenbach Falls,[11] both toppling over to apparent death. London erupted with dismay when FINA was published in *The Strand Magazine*. There would be no Sherlock Holmes stories for the next eight years. In EMPT, Holmes tells how Moriarty tracked him to Switzerland and describes their struggle at Reichenbach. Thus we learn how he came to survive. And, finally in VALL, Holmes, after cracking Fred Porlock's code (see section 3.6), suspects that Moriarty is behind the probable murder of John Douglas.[12]

Conan Doyle gives us some background on Professor Moriarty. We learn that he initially had great success as a mathematician. By age twenty-one, he had written a successful treatise on the binomial theorem. The expansion of the expression $[a + b]^n$ seems to have been first mentioned by Euclid around 300 B.C. A number of other mathematicians contributed to the concept, including Isaac Newton, who generalized the expression for fractional and negative values of n. Finally, in the 1820s, the Norwegian Niels Henrik Abel gave a proof for all values of n (Anderson 1989, 278). Note that all of this occurred well before Moriarty's time. So what was Moriarty doing with this old problem? Whatever it was, we are told in FINA that his treatise enjoyed a "European vogue." Its success led to Moriarty earning a chair in mathematics at a small university in England.

Later, in VALL, we learn that Moriarty also published the mathematically difficult "Dynamics of an Asteroid." Some Holmesians, ignoring the word "an" in the title, have speculated that Moriarty's work dealt with a general approach to the motions of all asteroids. The famous chemist/author Isaac Asimov, an avid Sherlockian, felt that

[11] Multiple plaques mark the spot.

[12] He disappeared at sea.

the title did imply that Professor Moriarty was discussing one particular asteroid. He claimed that Moriarty's book dealt with the motions of the planet that shattered to create the asteroid belt between Mars and Jupiter (see section 5.4). This clearly is a stretch since the title of the book states that the work is about an asteroid and not a planet. Calling it an "asteroid planet" (Schaefer 1993, 10) is not persuasive.[13]

With his academic career off to such a good start, it is puzzling why the professor turned to crime. But "dark rumors" caused him to resign his chair at the university. He became an "army coach" in London, hardly a high-paying job. Yet he had funds in six different banks (VALL). Eventually Holmes determines that he is at the center of a vast organization that controls most of the crime in London. Moriarty's organization deals in forgery, robbery, and even murder. Part of Holmes's evidence for Moriarty's illicit doings is the fact that he owns a painting by Greuze.

Jean Baptist Greuze (1725–1805) was a French painter who specialized in the "sentimental narrative in art" (Tansy and Kleiner 1996, 902). His most famous picture, *The Village Bride* (1761), hangs in the Louvre. It drew large crowds when exhibited at the 1761 Salon de Paris (Tansy and Kleiner 1996, 903). Greuze's paintings became very popular again in the time of Conan Doyle, drawing record prices at auction (Doyle and Crowder 2011, 126). How could a professor making only £700 per year afford such an expensive work of art? Obviously he either stole it or had some other source of income.

It should be noted that in VALL, Holmes likened Moriarty to Jonathon Wild, a London crime lord hanged in 1725 (Smith 2011, 124). One of Wild's strategies was to return goods he had stolen to the original owner—and collect a "finder's" fee. But Conan Doyle himself reportedly identified the more contemporary Adam Worth as the model for Professor Moriarty (Macintyre 1997, 223). Like Wild and Moriarty, Worth also had an extensive network of London thieves. He was actually labeled the Napoleon of the criminal world by Sir Robert Anderson, the head of Scotland Yard's criminal investigation (Macintyre 1997).

Worth was born in Prussia in 1844 (Doyle and Crowder 2010, 131). His family moved to America when he was five, and he eventually became a union soldier. Although he survived the first major battle, at Bull Run, he was listed as killed in action. He took this as an opportunity to disappear. He now began a career of theft, initially in the Boston area. His major haul came in November 1869 when he robbed the largest bank in Boston, Boylston National Bank. Using the alias William A. Judson, Worth rented the building adjacent to the bank. Then, having calculated the position on the wall of the bank's steel safe, he drilled at night until he could remove the back of the safe. The haul was reported to be between

[13] My own take on this issue ("Moriarty Vindicated," *The Baker Street Journal* 33(1), 1983, 37) is that the professor's book dealt with one asteroid, the one that collided with the earth near Yucatan. One result of this collision was the extinction of the dinosaurs.

$150,000 and $200,000 (Macintyre 1997, 38). Conan Doyle uses a similar ploy in REDH where the crooks tunnel into a bank vault.

The bank hired the Pinkerton Agency to pursue Worth (Doyle and Crowder 2010, 131). Feeling the pressure, Worth fled to Europe. He then assumed the name Henry J. Raymond, possibly taken from the recently deceased founder and editor of *The New York Times* (Macintyre 1997, 40). He used this alias for the rest of his life. In London, he set up the crime network that got Conan Doyle's attention and earned him the title "Napoleon of the criminal world."

Worth's most sensational crime was his theft of a Thomas Gainsborough painting, *The Duchess of Devonshire*. The duchess was a woman of great beauty. Her sexual life has been termed "raunchy in the extreme" (Macintyre 1997, 90). She allowed her husband's mistress to live with them so they could all enjoy a ménage à trois. Public interest in her was piqued when her portrait was painted in 1787 by the famous Gainsborough. The portrait, which had an interesting history itself (Macintyre 1997, 62), came up for auction at Christies in May 1876. Art dealer William Agnew bought the painting for £10,605. It was, at the time, the highest price ever paid at auction for a portrait. Almost immediately Agnew sold the painting to J. S. Morgan. It was to be a present for his son, the wealthy industrialist J. P. Morgan. Before transfer took place, Worth broke into Agnew's gallery on May 27, 1876, and stole the famous work of art.

Immediately Worth began writing a series of letters to Agnew, offering to return the painting for a fee. Perhaps Worth, like Conan Doyle, was familiar with Jonathon Wild's tactics. His initial demand was for $25,000 (he was writing from America). Negotiations failed, and Worth would keep the *Duchess* for almost twenty-five years. In 1901, with the Pinkertons acting as intermediaries, Worth returned the painting. He reportedly received $25,000, the exact figure he had sought twenty-five years earlier (Smith 2011, 125; Macintyre 1997, 253). Other sources cite different amounts. Agnew soon sold *The Duchess of Devonshire* to the sixty-four-year-old J. P. Morgan. The Morgan family eventually put the painting up for auction at Sotheby's in London on July 13, 1994. It was sold for £265,500—to the Duke of Devonshire!

The similarities of the two master criminals, Moriarty and Worth, are obvious. Both were at the center of an extensive crime ring in London. Both possessed an expensive work of art by a prestigious painter. The title of the fictional Greuze painting owned by Professor Moriarty has been termed "one of Conan Doyle's most delicious puns" (Macintyre 1997, 225). The painting's title, *Jeune Fille a l'agneau,* means "Young Girl with a Lamb." But most Holmesians would quickly point out the resemblance of the French word *agneau* to the name of the dealer from whom Worth stole the *Duchess*: Agnew (Macintyre 1997, 225).

Professor Moriarty is another excellent example of a dual person in detective fiction. We saw in the section on Poe in chapter 1 that the dual nature of humanity was featured in his Dupin stories. Recently there has been a study of the duality, or the bi-part soul, described in early detective fiction (Craighill 2010). Craighill traces this duality from its inception in Poe's *Murders in the Rue Morgue* (1841), through Inspector Bucket in Charles Dickens's *Bleak House* (1853) and Sergeant Cuff in Wilkie Collins's *The Moonstone* (1868), to Robert Louis Stevenson's *Dr. Jekyll and Mr. Hyde* (1886). Moriarty, an obvious addition to Craighill's list, has his bad side, the criminal mastermind, based on master criminal Adam Worth. Moriarty the scientist, the good part of the man, is based on the astronomer Simon Newcomb.

In addition to his work on the binomial theorem and on asteroids, Moriarty was interested in the motions of other celestial bodies as well. In VALL, he explains eclipses to Inspector MacDonald. The Canadian-American astronomer Simon Newcomb had exactly the same interests as those attributed to Moriarty. Newcomb, born in Nova Scotia in 1835, spent his working career in the United States. He was appointed professor of mathematics and astronomy at the U. S. Naval Observatory. He stayed there until his retirement in 1897 (Dictionary of Canadian Biography Online). Starting in 1884, Newcomb also had a part-time appointment as a math professor at the then "small" Johns Hopkins University. But, like Moriarty, he was forced to resign, although there were no "dark rumors."

His early research work focused on two of Moriarty's interests. He wrote an early unpublished work on the binomial theorem, and his first published paper dealt with a method of dynamics (Schaefer 1993, 11). His obituary in *The Times* noted an early paper on the orbits of asteroids. In the 1860s, Newcomb published several papers on the dynamics of individual asteroids. He is honored by having asteroid 855, Newcombia, named in his honor. Newcomb led eclipse expeditions in the 1860s and 1870s (Schaefer 1993, 11). So both Moriarty and Newcomb were interested in the binomial theorem, the motions of asteroids, and eclipses. It has been noted that paragraphs describing the scientific careers of the two would be nearly identical (Schaefer 1993, 11).

It was the genius of Arthur Conan Doyle that was able to make Professor Moriarty, with so few appearances in the Holmes stories, such a vivid, malevolent character.

Section 2.4 Other Important Characters

MYCROFT HOLMES

> *Occasionally he* is *the British government.*
>
> —Sherlock Holmes, *The Bruce-Partington Plans*

It is in the twenty-fourth story, GREE, that we, along with Dr. Watson, learn that Sherlock Holmes has a brother seven years his senior. Remember that in the second story, SIGN, Watson informed Holmes that he had a brother. Now, a mere two stories before Conan Doyle kills off Sherlock Holmes in FINA, we get a glimpse of Mycroft Holmes (see figure 2.8). And what a treat it is! Mycroft Holmes is surely one of best characters in the Holmes tales.

Physically the brothers are very different. Mycroft is described as corpulent, heavily built, and massive. We've seen that Sherlock was full of energy when on a case. He roams all over London and much of England as well. Mycroft, Sherlock tells us, "has no ambition and no energy." He "would rather be considered wrong than take the trouble to prove himself right." Mycroft is so unsociable that he is one of the founders of the Diogenes Club, "the queerest club in London." Its members are the "most unclubable men in town." No member is allowed to even take notice of other members, let alone actually talk to them. In fact, there is no talking permitted at the Diogenes Club, except in the "Stranger's Room." There, members may chat quietly with guests.

Figure 2.8 Mycroft Holmes, one of Conan Doyle's most unique characters

One thing the Holmes brothers do have in common is amazing deductive abilities. Mycroft, it seems, has even better powers of observation than Sherlock. We saw him outdo Sherlock in the scene with the billiard marker (GREE) in section 1.3. When Holmes brings Watson to the Stranger's Room at the Diogenes Club to meet his brother, we quickly get another look at Mycroft's genius:

> "By the way, Sherlock, I expected to see you round last week to consult me over that Manor House case. I thought you might be a little out of your depth."
>
> "No, I solved it."
>
> "It was Adams, of course."
>
> "Yes, it was Adams."

What is this? Does the brilliant Sherlock Holmes need someone with a higher intellect to consult when the case is too difficult? Indeed he does, as Sherlock admits in GREE:

> Again and again I have taken a problem to him, and have received an explanation which has afterwards proved to be the correct one.

When Watson remarks that Holmes's talent for deduction is due to training, Sherlock declares there is some heredity in it and acknowledges that Mycroft is the superior intellect:

> "But how do you know it is hereditary?"
>
> "Because my brother Mycroft possesses it in a larger degree than I do."

Yet the two are more partners than competitors. When Sherlock disappears after pushing Professor Moriarty over the Reichenbach Falls, it is Mycroft alone who knows that Sherlock is still alive. Despite drawing a modest salary of only £450 per year, he is able to supply the money that allows Sherlock to travel to such places as Tibet, Persia, Mecca, and France during the ensuing "Great Hiatus." In all of the sixty tales, only Mycroft is on a first-name basis with his brother. Even long-time roommate Watson only addresses him as Holmes. This is so sacred to Sherlockians that author Laurie King,[14] in her ongoing series, does not even have Sherlock's wife Mary Russell call him anything but Holmes!

If Mycroft Holmes is so lazy, what does he do? It turns out that he is a government employee. When we first meet him in GREE, published in 1893,

[14] Laurie King has written approximately twelve stories featuring Sherlock Holmes and his wife Mary Russell.

we are told that he "audits the books in some of the government departments." Fifteen years later, in *The Bruce-Partington Plans* (BRUC), Mycroft is rather more vital to England. Now Sherlock claims that "occasionally he *is* the British government." And, "Again and again his word has decided the national policy." It seems that Mycroft's specialty is omniscience.

Despite his reputation for lethargy, in cases of extreme danger, Mycroft is capable of physical action. When Holmes is fleeing Professor Moriarty in FINA, he says to Watson,

> In the morning you will send for a hansom, desiring your man to take neither the first nor the second which may present itself.

The driver of the third hansom is Mycroft, who has overcome his indolence to help his brother in this emergency. In GREE, he rouses himself to make the effort to come, for the first time, to the Baker Street lodgings to seek Sherlock's help.

Arthur Conan Doyle is at his best when writing about Mycroft Holmes. The use of Mycroft as a first name remains a rarity. Conan Doyle most likely took it from the well-known cricketer William Mycroft (1841–1892). He played mainly for Derbyshire, but also for the Marylebone Cricket Club. Conan Doyle himself was a talented cricketer who also played for Marylebone (Miller 2008, 241). Mycroft is mentioned in only four stories, GREE, BRUC, FINA, and EMPT. Yet he is one of the most memorable of all the characters in the Canon. Proof of this is the frequency with which he appears in all kinds of spin-off works, such as movies and books. For example, in Robert A. Heinlein's *The Moon is a Harsh Mistress* (1966), there is an omniscient computer named Mycroft (Redmond 1993, 42). There has even been a suggestion that it was Mycroft and not Sherlock who was modeled after Poe's detective. Sherlock is just much more active than Dupin or Mycroft (Propp 1978). Most readers leave the Holmes stories wishing that Mycroft had appeared more often.

MRS. HUDSON

> *. . . a long suffering woman*
>
> —Dr. Watson, *The Dying Detective*

Holmes and Watson were fortunate to have Mrs. Hudson as their landlady at 221B Baker Street. She is never physically described, save as having a "stately" tread (STUD). So the matronly image that persists is due more to drawings and movies than to Conan Doyle. But, as she appears in about a quarter of the sixty stories, we do get a good picture of her.

Due to her position in the household, Mrs. Hudson was frequently a conduit through which both clients and Scotland Yard inspectors reached Sherlock Holmes.

In *The Three Garridebs* (3GAR), she brings the card of John Garrideb, a.k.a. Killer Evans, up to Holmes on a tray. In DANC, she brings Holmes a telegram from New York with important news about Abe Slaney, the most dangerous crook in Chicago. It was Slaney who was writing to Elsie Cubitt using the "dancing men" code (see section 3.6). Of course, Mrs. Hudson generally conducted visitors up the seventeen steps to Holmes's lodgings. A great variety of people came to Baker Street to see Holmes. Among them were some sailors in BLAC, Inspectors Gregson and Baynes in WIST, and Cecil Barker in VALL. The king of Bohemia also visited the Baker Street rooms. Probably the most important visitor that Mrs. Hudson introduced was Mary Morstan, in SIGN. Mary, of course, was to become Mrs. Watson.

Meals were served at Baker Street, so Mrs. Hudson is also shown in this role. In NAVA, Holmes says, "Her cuisine is a little limited." Offers like "green peas at 7:30" (3STU) make this criticism understandable. On the other hand, Holmes is willing to request dinner for two at the end of MAZA. She also serves curried chicken in NAVA and woodcock in BLUE. Holmes is actually enthusiastic about Mrs. Hudson's "excellent" breakfasts (BLAC and NAVA). Apparently her specialty was ham and eggs with toast and coffee (Starrett 1934, 100). Mrs. Hudson was rather forgiving of Holmes's answer when, in MAZA, she asked, "When will you be pleased to dine, Mr. Holmes?" The snippy answer was "7:30, the day after tomorrow."

Fortunately for Sherlock Holmes, tolerance was Mrs. Hudson's strong suit. We noted in section 1.3 that Conan Doyle deliberately made Sherlock eccentric in an attempt to increase his appeal to readers. Mrs. Hudson had to deal with some strange behaviors. Holmes was called "the very worst tenant in London" (DYIN). He was very untidy, kept his cigars in a coal-scuttle, and his tobacco in a Persian slipper. He kept correspondence "transfixed by a jackknife" to the center of the wooden mantelpiece. At one point, he attempted to honor the queen by shooting bullets into the wall so that the holes formed the letters VR for Victoria Regina (MUSG). Mrs. Hudson constantly had to deal with "throngs of singular and often undesirable characters" (DYIN). An atmosphere of violence and danger was not rare. On a number of occasions, his chemical experiments (see chapter 4) filled the rooms with malodorous vapors.

Mrs. Hudson, however, was not tolerant of the Baker Street Irregulars (BSI). They were a group of "street Arabs," a phrase used in Holmes's day to indicate homeless children (Doyle and Crowder 2010, 120). They appear in the first two stories, STUD and SIGN. Holmes pays them a pittance to gather information for him. He claims that the BSI can "go everywhere, see everything, overhear everybody" (SIGN). The reason they tended to be overlooked in London was that the city then had some 30,000 abandoned children. Many of them "lived by stealing" (Doyle and Crowder 2010, 121). Mrs. Hudson's distaste for the group is understandable. Despite Holmes's belief in them, their performance record is mediocre. The first time we see them, in STUD, they succeed in locating Jefferson Hope's cab. But in SIGN, they are asked to find Mordecai Smith's boat, the steam launch *Aurora*. They fail, and Holmes himself has to track it down. The dramatic boat chase down the Thames, pursuing Jonathon Small and Tonga, follows.

Mrs. Hudson expresses disgust over the BSI in STUD and dismay in SIGN. To her, they are a noisy mob of dirty kids. It's obvious that they are an unruly crowd. Not even Holmes, their benefactor, can control them. The first time we meet them, in STUD, Holmes tells their leader Wiggins that only he should come to him to report. There is no need for the whole noisy crowd to come up to his flat. In the very next adventure, SIGN, Holmes has to make the same point as twelve eager kids crowd into his room. Mrs. Hudson was undoubtedly glad when they disappeared after the second story. Only once in the remaining fifty-eight tales are they mentioned. In *The Crooked Man* (CROO), one of the BSI trails Henry Wood, the crooked man whose appearance caused his former commander, Col. Barclay, to faint. This time, they are not a problem: There is only one "street Arab," he doesn't come to Baker Street to bother Mrs. Hudson, and he is successful in his task. Perhaps the BSI improved with age.

Why didn't Mrs. Hudson evict Holmes, her problematic tenant? One reason might be the "princely payments" mentioned by Watson (DYIN). We're also told that she was in awe of him. And she was fond of him as well; we clearly see that in her concern for his health. "I am afraid for his health," she tells Watson. When the supposedly dead Holmes shows up at Baker Street in EMPT, Mrs. Hudson is thrown into "violent hysterics." She never had need of the professional services of her brilliant lodger, and she was never directly involved in a case. We may wish that Conan Doyle had written a tale about some problem of hers. In fact, he did write two stories that involve landladies. But, alas, these cases were not about Baker Street and did not involve Mrs. Hudson. Conan Doyle created a different landlady in each case, Mrs. Warren in *The Red Circle* (REDC) and Mrs. Merrilow in *The Veiled Lodger* (VEIL).

In just a few stories does Mrs. Hudson do something other than introduce someone or cook. In NAVA, she goes along with the joke Holmes wishes to play on Percy Phelps (see figure 2.9). She serves Mr. Phelps a covered dish, but with no food on the plate. Instead, Percy finds the recovered treaty under the lid. But Mrs. Hudson's crowning moment comes in EMPT. Here she plays a vital role, one that puts her in danger. We can imagine that she is pleased to be able to serve Holmes in this way. Col. Sebastian Moran, "the second most dangerous man in London" and formerly chief of staff to the late Professor Moriarty (EMPT), arrives in the vacant house opposite 221B Baker Street. His goal is to kill Holmes using his high-powered air gun. Moran was known to be "the best heavy-game shot that our Eastern Empire has ever produced" (EMPT). But Holmes has had a bust of his head made. To Moran, it seems that Holmes is visible through the window. But Holmes, along with Watson, is in

Figure 2.9 Mrs. Hudson helps Holmes play a joke on Percy Phelps in *The Naval Treaty*.

the vacant house too. As they wait, Watson is amazed to see that the bust has moved. Holmes responds, "Of course it has moved. Am I such a farcical bungler, Watson, that I should erect an obvious dummy." Mrs. Hudson has crawled to the bust every fifteen minutes to move it, placing herself in danger to assist Holmes. Then, after Moran fires and shatters the bust, Mrs. Hudson retrieves the spent bullet for evidence.

It is likely that when Holmes purchased a villa in Sussex with "a great view of the Channel" (*The Lion's Mane*, LION), Mrs. Hudson went along. He tells us that his "old housekeeper" lives there with him. Since we only know of him living on Baker Street, many have claimed that this is Mrs. Hudson. Because of her service and loyalty to Sherlock Holmes, Mrs. Hudson generally ranks as one of the favorite characters in the entire Canon. One scholar has done her the honor of referring to her as *the* woman, a title usually reserved for Irene Adler of STUD (Cooke 2005).

SCOTLAND YARD

Now, in my opinion, Dupin was a very inferior fellow.

—Sherlock Holmes, *A Study in Scarlet*

Conan Doyle's aim was to make the brilliance of Holmes the selling point in his stories. So in the first story (STUD), he has Holmes criticize his fictional predecessors Dupin and Lecoq. Then throughout the stories, he makes Watson the unaware chronicler. As Holmes says in BLAN,

> one to whom each development comes as a perpetual surprise and to whom the future is always a closed book, is indeed an ideal helpmate.

Then in addition to contrasting Holmes with Dupin, Lecoq, and Watson, Conan Doyle also uses the official police force to reinforce the idea of Holmes the genius. Right from the start, in STUD, Holmes is at odds with Scotland Yard Inspectors Lestrade and Gregson. When he arrives at the crime scene, Holmes complains that the police have obscured clues on the pathway. "If a herd of buffaloes had passed along, there could not be a greater mess."[15] Then Holmes tells Watson that he is superior to the Scotland Yarders and bemoans the fact that they will take the credit after he solves the case. Holmes actually laughs at Inspector Lestrade's conclusion that the "RACHE" written on the wall in blood refers to a woman named Rachel. Still Holmes is willing to assist them by giving the description of the culprit cited earlier. But as we've seen, he can't resist that parting shot about not wasting time looking for the nonexistent Rachel.

Things go much the same way in the second story, SIGN. Inspector Athelny Jones denigrates Holmes's methods (Doyle and Crowder 2010, 113) and tells him there is no room for theories in the case, because "facts are better than theories." Jones comes to an erroneous conclusion and arrests the wrong man. Yet, at the end of the story, Watson remarks that he has found a wife, Jones has gotten the credit, and Holmes has gotten nothing. That is when Sherlock Holmes returns to the "cocaine-bottle."

In the fourth story, REDH, Inspector Peter Jones is dismissive of Holmes's methods. And Holmes refers to Jones as an imbecile. In the sixth story, BOSC, Holmes calls Lestrade an imbecile, and Lestrade tells Holmes that he is ashamed of him. Lestrade accuses the wrong man of the murder of McCarthy. Holmes, believing that Turner was justified in killing the blackmailing McCarthy, decides to give Lestrade only a detailed description of the murderer instead of his name:

> ...a tall man, left-handed, limps with the right leg, wears thick-soled shooting boots and a gray cloak, smokes Indian cigars, uses a cigar holder, and carries a blunt pen-knife in his pocket.

The befuddled detective is slow to catch on:

> LESTRADE: Who was the criminal, then?
>
> HOLMES: The gentleman I describe.

[15] Holmes uses the same phrase in *The Boscombe Valley Mystery*, the sixth story.

> LESTRADE: I really cannot undertake to go about the country looking for a left-handed man with a game leg.

Twenty-one different Scotland Yard detectives appear in forty-two of the sixty stories (Doyle and Crowder 2010, 107). We find Lestrade in fourteen stories, Gregson in five, and Stanley Hopkins in four. Most of the others are in only one story. As time goes on, Holmes's profession as the world's only "consulting detective" becomes more established. The Scotland Yard inspectors become convinced that he is an ally. Thus we find that the relationship mellows, and Holmes no longer frets about not getting credit. In *The Cardboard Box* (CARD), the sixteenth story, he says, "I'd prefer not to be mentioned." In NAVA, number twenty-five, Inspector Forbes of Scotland Yard accuses Holmes of seeking credit at the expense of the Yard. Holmes responds,

> On the contrary, out of my last fifty-three cases my name has only appeared in four, and the police have had all the credit in forty-nine.

Forbes immediately changes his approach and says,

> I'd be very glad of a hint or two.

The verbal insults stop too. In NORW, Lestrade admits that "we owe you a good turn at Scotland Yard." By the time of SIXN, Lestrade has become a friend (see figure 2.10). Watson records, "It was no very unusual thing for Mr. Lestrade, of Scotland Yard, to look in upon us of an evening, and his visits were welcome to Sherlock Holmes. . . . " At the conclusion of the case, Lestrade says,

> We're not jealous of you at Scotland Yard. No, sir, we're very proud of you, and if you come down tomorrow, there's not a man, from the oldest inspector to the youngest constable, who wouldn't be glad to shake you by the hand.

Holmes's comments about the Scotland Yard detectives become kinder with time. In HOUN, he calls Lestrade "the best of the professionals." In BRUC, he praises one of Lestrade's observations: "Good Lestrade, very good." In CARD, he praises Lestrade's tenacity. In REDC, Watson talks of Gregson's courage.

In the later stories, Scotland Yard is sending business Holmes's way. In MISS, Stanley Hopkins advises Cyril Overton to consult Holmes. In RETI, the Yard refers Josiah Amberly to Holmes. Holmes, in turn, is willing to point the police in the right direction. For example, in VALL, he advises Inspectors MacDonald and Mason to forget about the cyclist. In BLAC, he suggests that Hopkins should

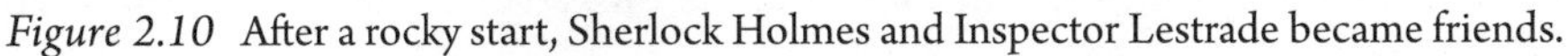

Figure 2.10 After a rocky start, Sherlock Holmes and Inspector Lestrade became friends.

focus on the tobacco pouch. The only time he refuses to help is in CHAS. Lestrade is seeking the two men who were seen, and almost caught, fleeing the scene of Milverton's murder. The two men, of course, were Holmes and Watson.

At the start, the Scotland Yard detectives were used as a contrast to Holmes's deductive approach. As Holmes became more closely associated in the public mind with brilliant reasoning, Conan Doyle let his relationship with the official force change. He no longer needed them to be fools. The whole tone of the Holmes/Scotland Yard interaction evolved into something much more realistic.

3

Sherlock Holmes

Pioneer in Forensic Science

Section 3.1 The Methods of Bertillon

His conversation, I remember, was about the Bertillon system of measurements, and he expressed his enthusiastic admiration....

—Dr. Watson, *The Naval Treaty*

Holmes may have admired Bertillon's work, but that did not prevent him from being resentful about it in *The Hound of the Baskervilles* (HOUN). When Dr. James Mortimer told Holmes that Bertillon was the highest expert in Europe, Holmes admitted that he was offended by the ranking. So who was this man held in such high regard?

Alphonse Bertillon was a French anthropologist born in 1853. His poor academic performance was followed by difficulty holding a job. In 1879, his influential father Louis, a famous physician and anthropologist, obtained a job for him as a clerk with the Parisian police (Wagner 2006, 97–98). He started work in March 1879, and became interested in the problem of identifying recidivists,[1] that is, repeat offenders. It was French policy to exile recidivists to their colonies (Cole 2001, 33). But there was no procedure for identifying them. Fingerprinting did not exist, and even mug shots were not yet used. Upon a second arrest, recidivists would merely use a pseudonym. Bertillon wanted to develop a system of identification based on ideas mentioned in 1840 by a Belgian statistician named Quetelet (Wagner 2006, 98). Bertillon found his job with the police to be very boring, as he collected and filed much information, most of it never used again and worthless. So, on October 1, 1879 (Cole 2001, 49), he submitted a report

[1] The word "recidivism" entered the English language in 1886 (Cole 2001, 53).

proposing a method of identification using body measurements. The report was ignored (Wagner 2006, 98).

Louis Bertillon liked his son's suggestion. Louis had in fact attempted to classify people, not identify them, by measuring the lengths of their bones. So he was naturally attracted to Alphonse's idea to use such measurements to identify criminals (Cole 2001, 34). In 1882, with help from his influential father, Alphonse Bertillon was given two assistants and some funding. He was given three months to identify a repeat offender. He succeeded with one week remaining. A man convicted of a crime and using the name DuPont was found to have the exact same measurements as a previously convicted man named Martin. Confronted with the evidence, DuPont confessed.

The measurement system, or anthropometry, that Bertillon devised at this point had three parts. The first part involved eleven body measurements using calipers, each measurement being done three times and averaged.[2] The second part consisted of a precise physical description of the person with emphasis on the ear. Finally, any peculiar marks on the body were recorded. Two photos completed the characterization of the prisoner (Cole 2001, 37).

By 1880, the Paris police had 75,000 photos of criminals, catalogued in alphabetical order. This unproductive arrangement proved to be entirely unsatisfactory. Bertillon's method was used to arrange cards according to his measurements. Soon there were 120,000 cards in groups of about twelve (Cole 2001, 45). Now the system could actually locate and describe people. In the first full year of using Bertillon's system, 1884, Paris police identified 241 recidivists (Cole 2001, 49). Continued success led, in 1888, to the formation of the Department of Judicial Identity, with Alphonse Bertillon as its head. The system, referred to as Bertillonage, swept the world. The United States adopted it in 1887 and Great Britain did so in 1894 (Cole 2001, 51).

In the meantime, the use of fingerprints as a means of identification was taking hold as well. Bertillon resisted the use of fingerprints, though not totally (Wagner 2006, 105). In fact, Bertillon was the first European, in October 1902, to solve a murder using fingerprints. There ensued a forty-year battle between Bertillonage and fingerprinting for ascendancy in identification (Cole 2001, 32). Bertillon's errors in the Dreyfus case in 1894 and the theft of the Mona Lisa in 1911 were factors in the ultimate demise of his method.

Bertillon's confused testimony about handwriting was a definite factor in the conviction of Alfred Dreyfus in 1894 (Wagner 2006, 163). Dreyfus was accused of treason for giving French military secrets to the Germans. He was sent to the penal colony at Devil's Island. The author Emile Zola published a public letter, "J'accuse," to French President Félix Faure. Zola's spirited defense of Dreyfus was a big factor in shaping French public opinion in Dreyfus's favor.

[2] For a list of the eleven measurements, see Wagner 2006, 98 or Cole 2001, 37.

Because the evidence against him was flimsy, an outcry resulted in a retrial in 1899. Most were amazed that Dreyfus was again found guilty. But he was eventually exonerated in 1906 and reinstated. The case was damaging to Bertillon because he had stepped outside his area of expertise. He had injured his reputation, as nearly all the French people came to believe that Dreyfus was innocent.

Then, when the Mona Lisa was stolen in August 1911, Bertillon suffered another humiliation. Surely he would be able to match the fingerprint of the thief's left thumb found on the glass case in which the Mona Lisa was displayed. But he failed to do so. The thief, Vincenzo Perugia, had an arrest record in France. Bertillon did have a print—but only of his right thumb (Wagner 2006, 105). These setbacks were the beginning of the demise of Bertillonage. Ultimately fingerprints proved to be more reliable than Bertillon's system. Alphonse Bertillon died in 1913.

Section 3.2 Fingerprints

You are aware that no two thumb-marks are alike?

—Inspector Lestrade, *The Norwood Builder*

Despite Sherlock Holmes's "enthusiastic admiration" for Bertillon's system, he never used it. He did, however, make use of fingerprints. They are mentioned in seven of the sixty Sherlock Holmes tales.

A BRIEF HISTORY OF FINGERPRINTING

It is known that very long ago the Chinese considered the impression of a fingerprint on a document to be a unique signature. They were taken as identifying seals on Chinese bills of sale in the third century B.C. About 2000 B.C., fingerprints were used by Babylonians to seal contracts (Bigelow 1957). Modern use of fingerprinting may have begun with Govard Bidloo, a Dutchman, and Marcello Malpighi, a professor of anatomy at the University of Bologna (Kaye 1995, 13). In 1685 and 1687, respectively, they recognized the importance of fingerprints. The English engraver Thomas Bewick in 1804 and 1818 made wood engravings of the patterns of his fingerprints for use as his trademark (Kaye 1995, 13).

The Tichborne case in England in the 1870s brought great attention to the need for a reliable identification system. Arthur Orton from Australia claimed to be the British heir Roger Tichborne, missing at sea for over ten years. It took three years to settle the case, which generated much publicity and resulted in a feeling that a faster method of identification was needed. Prior to the use of fingerprints, identity was established by letters of reference, official papers, and photographs. We've seen that use of the system known as Bertillonage or anthropometry preceded fingerprinting in criminal detection.

The use of fingerprints to identify criminals in Britain, and eventually much of the world, can be traced to a letter to the editor of *Nature*, dated October 28, 1880. It was written by Henry Faulds, a Scottish medical missionary at Tsukiji Hospital in Tokyo. There a thief had left a fingerprint on a wall. It did not match the print of the main suspect. It did match another suspect, who then confessed. Faulds noted that monkeys have fingerprints that are similar to humans. He claimed that heredity plays a role in influencing the form of fingerprints. He described one of the common features of fingerprints by means of a term that is still used today, whorl. He remarked that fingerprints might be useful in identifying criminals and noted that he had knowledge of two cases of such use. Faulds also made the important assertion that fingerprints are unchanged throughout one's life, calling them "the for-ever-unchangeable finger-furrows" (Wagner 2006, 102). He even whimsically pointed out that when Dr. Jekyll transformed himself into Mr. Hyde, his fingerprints would remain unchanged (Cole 2001, 3). Henry Faulds would eventually argue against the idea that no two fingerprints are alike (Cole 2001, 188).

A response to Faulds's letter was published in the November 25, 1880, issue of *Nature* by W. J. Herschel, who, as a British official in Bengal, India, reported that he had been taking fingerprints there for more than twenty years. He had started in 1860 in order to identify government pensioners. Some were showing up twice to collect their pensions. As soon as Herschel began fingerprinting as a form of identification, the attempts at double collection ceased. He then extended the practice to criminals in jail. Herschel disagreed with Faulds's idea that fingerprints could be use to suggest ethnicity or genetic relationship. He had observed wide differences in fingerprints within families. He did not believe that fingerprints could distinguish ethnicity or sex.

In 1880, Faulds wrote a letter about his fingerprint work to Charles Darwin. Darwin forwarded the letter to his cousin Francis Galton (Cole 2001, 74). Galton, impressed by the discussion of fingerprints, asked the editor of *Nature* for the address of the discoverer of fingerprinting and was given Herschel's name. He visited Herschel, who willingly handed over all of his materials. A German anatomist named J. C. A. Mayer claimed in 1788 that a person's fingerprints were unique. Galton, in a three-year study, proceeded to verify Mayer's claim (Klinger 2006, 207; Bigelow 1957, 91). In the early 1900s, an article in *Scientific American* reported that the probability of two fingerprints being alike was 1 in 10^{60} (Cole 2001, 177). For all practical purposes, this is a probability of zero. The uniqueness of fingerprints is still today of immense importance in criminal identification. Galton then made an extensive collection of fingerprints in the late 1800s. He was initially studying inheritance, but eventually wrote the first textbook on fingerprints, asserting that they are never duplicated and remain unchanged for life (Klinger 2006, 207). He even made repeat measurements of one person's fingerprints over a period of fifty years. After over one hundred years of unchallenged use in the courtroom, fingerprints have recently come under renewed scrutiny. Galton's conclusions have been questioned. Has their uniqueness been sufficiently tested? Should a study be done to put fingerprinting on a firmer basis (Cho 2002; Specter 2002)? On January 7, 2002, Judge Louis H. Pollak, a former dean of law schools at Yale and the University of Pennsylvania, issued a ruling limiting the use of fingerprints in a murder case in Philadelphia. Then, on March 13, 2002, Judge Pollak vacated that order, and fingerprints were allowed. So far, this seems to be the end of that battle.

In 1892, Galton's influential book entitled *Finger Prints* led to the establishment of a committee to consider the advisability of adopting fingerprinting as a method of identifying criminals. The committee's system of classification of fingerprints, adopted in 1901, was known as the Henry system after committee member Sir Edward Richard Henry, who was later director of Scotland Yard (Kaye 1995, 14). At the time, Henry was a British civil servant in Calcutta, and he made substantial contributions to the method of classifying fingerprints. In July 1897, he persuaded the governor-general of India to adopt fingerprints as the sole means of identifying criminals. By August 1897, Henry had solved a number of crimes using fingerprints and in 1900, he published his system. Henry's work was greeted with such acclaim that he was appointed commissioner at Scotland Yard on May 31, 1901. By July 1901, he had instituted the Central Fingerprint Branch. In 1905, the Stratton case became the first instance of conviction in England for murder based on fingerprint evidence (Rennison 2005, 224). By 1910, the Henry system had been adopted throughout Europe. Despite this, Oscar Slater was wrongfully convicted in 1909 of the murder of Marion Gilchrist, even though a bloody hand print had been left on a chair at the scene of the murder. So, as late as 1909, Scotland Yard was not totally using fingerprinting. Conan Doyle had been personally involved in demonstrating Slater's innocence (Miller 2008, 292) by publishing *The Case of Oscar Slater*. But his demand for a new trial was denied, and Slater was imprisoned for eighteen years.

In the United States, the International Association of Chiefs of Police started fingerprint files in 1896. New York state authorities began collecting fingerprints of prisoners in 1903; but it was not until 1928 that New York required all offenders to be fingerprinted. On November 2, 1904, the warden of the U.S. Penitentiary at Leavenworth, Kansas, was authorized to take fingerprints of federal prisoners. In 1911, the Supreme Court of the State of Illinois upheld the legality of the use of fingerprints for identification of criminals. By the early 1970s, U.S. security authorities had over 200 million fingerprints on file. Eventually the FBI was receiving thousands of fingerprint requests each day.

However, in the early days of using fingerprints for identification, the near futility of finding a match by means of a manual search of existing fingerprint files proved to be a tremendous hindrance. With millions of fingerprints on file, the time required to find the right print was enormous. In addition, the recorded prints were often of mediocre or poor quality. The development of the Automated Fingerprint Identification System (AFIS) made fingerprint identification much swifter and therefore more useful. The success rate in identifying criminals increased by a factor of five when AFIS supplanted manual searches.

Much of the credit for the change is due to a San Francisco police inspector named Ken Moses. He was enraged by the 1978 murder of a forty-seven-year-old San Francisco woman who had survived the Nazi concentration camps. The only evidence was a set of three fingerprints left on an upstairs windowsill. Moses was faced with the task of matching these prints with those of 400,000 people taken in San Francisco over forty-five years. He began in 1978 and six years later was still hunting whenever he could find time away from his other assignments. Back in 1978, Moses had read about computerized fingerprint identification systems. He requested such a system and succeeded in getting his request put into the departmental budget. It was not purchased, however, due to budget restraints. Moses then received permission to attempt to raise the money in the community. His efforts, which involved lecturing to civic groups on the issue, did not succeed in raising money, but they did increase community awareness. When his group was able to get the issue on the ballot in 1982, it passed with 80 percent approval, and funds became available. In 1984, San Francisco's AFIS went online. Moses had a match within sixty seconds of entering his prints into AFIS. Two days later, the killer was arrested and in 1985 pleaded guilty to first-degree murder (Fincher 1989, 201).

HOLMES'S USE OF FINGERPRINTING

In the Sherlock Holmes stories, there are several cases where a fingerprint is noted but isn't used to apprehend anyone. The first of these is *The Sign of the Four* (SIGN). Holmes notes that a thumb mark was on the envelope mailed by Thaddeus Sholto to Mary Morstan. Holmes suspects that it was made by the postman. It turns out that he does not need to investigate the print because Sholto reveals his identity to Mary Morstan. In *The Man with the Twisted Lip* (TWIS), there is a greasy thumbprint on the envelope containing the note from Neville

St. Clair to his wife. It is eventually of no use as it belongs to his acquaintance who posts the letter. Meanwhile, Holmes solves the case by other means. In *The Cardboard Box* (CARD), Holmes notes two "distinctive" thumb marks on the cardboard box sent by Jim Browner to Susan Cushing. These prints are not used, as again Holmes solves the case by other means. In another story, *The Three Gables* (3GAB), the inspector assigned to this case keeps a page from Douglas Maberly's novel because it may have prints on it. In all of these stories, Holmes and the official police are looking for fingerprint evidence, but no helpful prints are found.

There are two cases where the absence of fingerprints is noted by Holmes. In *The Three Students* (3STU), Holmes observes that there are no fingerprints on the proofs of Hilton Soames's exam papers. In *The Red Circle* (REDC), a corner has been torn from the instructions sent to Mrs. Warren. Holmes surmises that it was done to eliminate a print.

In *The Norwood Builder* (NORW), Inspector Lestrade finds the thumbprint of the main suspect, John Hector McFarlane, on a wall in blood. He triumphantly asks Holmes,

> You are aware that no two thumb-marks are alike?

Holmes is indeed aware of the uniqueness of fingerprints, but he knows that the print was placed there *after* McFarlane had been taken into custody. Only Holmes had done a thorough examination of the wall the day before. The print had been placed there during the night by Jonas Oldacre in order to incriminate John Hector McFarlane. Oldacre had obtained it from Macfarlane by having him press a wax seal on his will. Jonas Oldacre also must have known about fingerprints. Otherwise he would have placed anyone's print on the wall and not bothered to get McFarlane's in wax.

This use of the thumbprint in NORW may have occurred to Conan Doyle upon his reading in the June 27, 1903, issue of the magazine *Tit-Bits* an article entitled "Criminals Convict Themselves." That article reports on a case in Yorkshire where a burglar took the time to look at a book and left a dirty thumbprint in it. NORW was published in November 1903 (Edwards 1993, *The Return of Sherlock Holmes*, 338). It is likely that this is the first time in literature that the idea of a false fingerprint is used.

SUMMARY

Arthur Conan Doyle was one of the very first authors to make use of the emerging technique of identification by fingerprints. The publication of NORW in 1903 preceded by two years the first successful use of fingerprints by the police.

The Stratton case of 1905 was the first time a murderer had been convicted on the basis of fingerprints (Rennison 2005, 224). But Mark Twain's writings referred to fingerprints even earlier than did those of Conan Doyle. Twain first mentions prints in *Life on the Mississippi* (1883). Here a man identifies the killers of his wife and child using a bloody thumbprint. Again in *Pudd'nhead Wilson* (1894), Twain makes very extensive use of fingerprints for identification, well before agencies had adopted the method. In this story, Wilson is considered an eccentric partly because of his hobby of repeatedly collecting the fingerprints of everyone in Dawson's Landing, a town on the Mississippi river. But Wilson is able to use his fingerprint collection to show that his Italian clients are innocent of a murder charge. In a more significant episode, Wilson also proves that the slave baby Chambers was exchanged in the cradle with the master's son Tom. Thus the real Tom was raised as a slave. Meanwhile, Chambers, only fractionally black and resembling his half-brother, became the heir. Wilson's fingerprint collection corrects all of this. One of the main points of Twain's story is that the slave-baby-turned-master treats the black slaves cruelly.

Conan Doyle's seven references to fingerprinting in the Holmes stories is an indication of the voracious reading habit that kept him so well informed.[3] By the time Scotland Yard had adopted fingerprinting for identification in 1901, Conan Doyle had already written three stories mentioning the method. He would make it the centerpiece of another Holmes tale, NORW, published in 1903. His preference for fingerprints over the Bertillonage methods (seven mentions versus two) shows him to be on the side of the winner in the forty-year competition between the two.

Section 3.3 Footprints

> *Footprints? Yes, footprints. A man's or a woman's?*
> *Mr. Holmes, they were the footprints of a gigantic hound.*
>
> —*The Hound of the Baskervilles*

INTRODUCTION

In the first Holmes story, *A Study in Scarlet* (STUD), published in 1887, we see that Sherlock was already using footprints in his work. His description

[3] For a contrasting view, see Paige 2002, 39.

of Constable Rance's movements is so precise that Rance blurts out, "Where was you hid to see all that." In *The Lion's Mane* (LION), published in 1926, Holmes is still using footprints. Here he observes that Fitzroy McPherson's footsteps are the only ones on the path down to the beach. So, for over forty years Conan Doyle has Sherlock Holmes use footprints in his investigations. Footprints were held in such high regard as a forensic tool that in about 1890, a letter to *The Times* suggested that fingerprints might be *almost* as good as footprints (Fido 1998, 89).

A BRIEF HISTORY OF FOOTPRINTING

In the Book of Daniel (part of the Bible or the Apocrypha, depending on one's personal beliefs), the king of Persia leaves large quantities of food out each night for the idol Bel (or Baal). The priests of Bel have persuaded the king that Bel comes every night and takes the food. But Daniel spreads ashes on the floor one night and is able to show the king that the priests themselves are taking the food. Despite this early example of the use of footprints, a "science" of footprinting for identification never developed (Moenssens et al. 1995, 614).

In the 1980s, Professor Louise Robbins from the University of North Carolina, Greensboro, tried to put footprint analysis on a scientific basis, publishing a book aimed at that goal (Robbins 1985). She served as an expert witness on footprints in at least twenty trials. Her testimony was a factor in several cases when she was able to testify to identities even when the state's own crime laboratories where unable to confirm them. Her methods, fully described in her book, have since been severely criticized (Moenssens et al. 1995, 619). One website, referring to her as an "infamous charlatan," reports that the American Academy of Forensic Sciences decided in 1987 that her work had no scientific basis (Zerwick 2011). A law professor described her work in this way: "It barely rises to the dignity of nonsense." So, although footprints are not as definitive as fingerprints, they are "probably the oldest of all detection techniques" (Fisher 1995, 277).

An interesting case of footprint evidence occurred in the O. J. Simpson criminal and civil trials. Footprints at the crime scene were shown to be of size-twelve Bruno Magli shoes, Lorenzo style. At his 1994 criminal trial, Simpson denied ever owning a pair of these somewhat rare, expensive shoes. The prosecution was not able to demonstrate that he did own Bruno Magli shoes. But by the time of his subsequent civil trial, a number of photographs showing him wearing such shoes on September 26, 1993, were found. These photographs were admitted into evidence at the civil trial. Mr. Simpson was acquitted at the criminal trial and convicted at the civil trial (Murray). This is not to say that the footprint evidence was *the* determining factor in the different verdicts, but it may have had some impact.

Apparently the "Unabomber," Ted Kaczynski, was concerned about footprint evidence. Fox News reported on November 29, 2006, that he had affixed smaller soles to the bottom of a pair of shoes found in his Montana cabin. His hope was to use these shoes to evade authorities who were chasing him. Foot impressions have long been permitted as evidence on a case-by-case basis. Sherlock Holmes made good use of footprints in his detective work, as shown in the next section.

HOLMES'S USE OF FOOTPRINTS

In solving his cases, Sherlock Holmes made more use of footprints than fingerprints. We've seen that the then-emerging science of fingerprinting for identification is mentioned in only seven of the sixty Holmes tales. Footprints, however, are mentioned in twenty-six of the sixty cases (Tracy 1977, 128). Clearly they were one of his major investigative tools. Conan Doyle chooses a number of different materials upon which footprints are left: clay soil (STUD), snow (*The Beryl Coronet,* BERY), carpet (*The Resident Patient,* RESI), dust (Tonga in SIGN), mud (Jonathon Small in SIGN), blood (REDC), a curtain (the mongoose in CROO), and ashes (GOLD) (Vatza 1987, 17). Holmes uses all of them as evidence.

Here are a few cases where Holmes produces some results using footprints. In the first two tales, STUD and SIGN, Holmes is able to trace movements of people so accurately that he startles them. In SIGN, just as in STUD, the movements of Jonathon Small and Tonga are as accurately determined by Holmes as were John Rance's in STUD. In fact, Jonathon Small is so surprised by Holmes's comments that he says, "You seem to know as much about it as if you were there." Of course Holmes is aided here by the fact that one of them, Tonga, is a pygmy from the Andaman Islands and the other has a wooden peg leg. When Holmes reveals Tonga's footprints on the dusty floor, the shocked Watson says, "Holmes, a child has done this horrid thing." Holmes, of course, has a different interpretation.

Holmes's luck in SIGN, to have two such distinctive footprints, is repeated in BERY. Here four people leave footprints: Sir George Burnwell's boot, Lucy Parr's shoe, Arthur Holder's naked foot, and Francis Prosper's wooden leg. Holmes declares, "A very long and complex story was written in the snow in front of me." Holmes proceeds to sort out the movements by these footprints left in the snow. He observes that both Lucy Parr and Arthur Holder ran. The servant girl, Lucy Parr, was meeting her beau, Francis Prosper. She ran when discovered. They have nothing to do with the theft of the beryl coronet. Sir George Burnwell stole the jeweled coronet and Arthur Holder hastily chased him. The innocence of the accused Arthur Holder is established by Holmes's reconstructing the crime scene based largely upon footprint evidence. This idea of running footprints is seen again in the opening scenes of HOUN. In addition to the footprints of a gigantic hound, the yew alley at Baskerville Hall had the footprints of Sir Charles Baskerville. Holmes deduces that the change in Sir Charles's footprints halfway down the alley was due to his running from the hound, not tiptoeing as had been suggested.[4]

In *The Devil's Foot* (DEVI), Holmes once again encounters two footprints of very different characters. Holmes is able to distinguish the normal footprint of Mortimer Tregennis from the ribbed tennis shoe worn by Leon Sterndale. Footprints are part of the evidence that enables Holmes to deduce that Mortimer Tregennis murdered his sister Brenda. In revenge for the loss of his secret love, Dr. Leon Sterndale causes the death of Mortimer Tregennis. This case is one of several in which Holmes solves the crime but allows the culprit to go free. He decides that Sterndale's actions are justified.

In RESI, he amazes Watson by using footprints on the carpet to deduce the order in which the perpetrators ascended the stairs. Holmes then proceeds to precisely describe their movements in Mr. Blessington's room before they hanged him.

There are two cases wherein Holmes acts to "develop" a footprint. In DEVI, Holmes kicks over a water pot in order to get an impression of Mortimer Tregennis's shoeprint. In GOLD, borrowing from the Book of Daniel, he uses tobacco ashes dropped on a carpet to show the presence of Anna Coram. His furious smoking of cigarettes makes him seem nervous and embarrasses Watson, who doesn't realize what Holmes is trying to do. When Holmes leaves and quickly returns to Professor Coram's room, he is able to see Anna's footprints in the ashes. In this manner, she is forced to emerge from her hideout. It is notable that the absence of Anna's footprints on the path outside is also part of Holmes's evidence. That absence is why he suspects and then proves that she is concealed behind the hinged bookcase in her husband's chamber. The absence of footprints is a factor in other cases as well: *Black Peter* (BLAC), *The Five Orange Pips* (FIVE), *The Reigate Squires* (REIG), *The Naval Treaty* (NAVA), and 3STU (Tracy 1977, 129).

The *Boscombe Valley Mystery* (BOSC) is one case that Holmes solves almost entirely by use of footprints. In this case, John Turner has killed the

[4] Sir Charles's death by heart attack while fleeing the hound of the Baskervilles down the yew alley gave rise to a medical term. The "Baskerville Effect" refers to heart attacks brought on by extreme emotional stress. Conan Doyle first describes a death by heart attack due to fear in SIGN. There Captain Morstan, Mary Morstan's father, undergoes a similar fate. (See Phillips et al. 2001, 1443–1446.)

Figure 3.1 Holmes's use of footprints in *The Boscombe Valley Mystery* was particularly effective.

blackmailing Charles McCarthy. But suspicion has fallen on McCarthy's son James. The McCarthys' maid has provided Holmes with boots belonging to both McCarthys (see figure 3.1). After measuring them very precisely, Holmes heads to Boscombe Pool, the scene of the murder. Once there, he must deal with the extraneous footprints of a number of people, including Inspector Lestrade.

> HOLMES: What did you go into the pool for?
> LESTRADE: I fished about with a rake. I thought there might be some weapon or other trace. But how on earth…
> HOLMES: Oh, tut, tut, I have no time. That left foot of yours with its inward twist is all over the place.

Holmes is able to trace the movements of both McCarthys, as well as Turner. He deduces that James McCarthy left three sets of footprints, one when he was running. This agrees entirely with his contention that he ran back to help his father after hearing an outcry. Persuaded that the younger McCarthy did not kill his father, Holmes uses the third distinct type of footprint to gather evidence against the true killer, John Turner. "What have we here? Tiptoes, tiptoes. Square too, quite unusual boots. They come, they go, they come again." Turner had to return to retrieve a cloak he had left behind at the scene.

Lestrade, contemptuous of Holmes's theories and still convinced that James McCarthy killed his father after an argument, is reluctant to accept Holmes's description of the murderer cited in chapter 2:

> ...a tall man, left-handed, limps with the right leg, wears thick-soled shooting boots and a gray cloak, smokes Indian cigars, uses a cigar holder, and carries a blunt pen-knife in his pocket.

Despite all of this information, Lestrade fails to capture Turner. This is one of the cases where Holmes decides, after giving Lestrade the above detailed clue, to let the culprit go free. He is in sympathy with the terminally ill Turner.

One last set of footprints must be mentioned. In *Charles Augustus Milverton* (CHAS), Inspector Lestrade finds footprints outside the residence of Charles Augustus Milverton, "the worst man in London." Little does he know that the prints belong to Holmes and Watson who were inside and witnessed the murder of Milverton. It is another case in which Holmes lets the perpetrator, this time Lady Eva Brackwell, go free.

SUMMARY

It is a tribute to Sherlock Holmes that he was able to so brilliantly solve crimes despite the lack of modern methods. His forty-year use of footprints is particularly impressive. He got so adept at reading them that in *Wisteria Lodge* (WIST), Holmes was able to tell the size of a print at a glance: "A number twelve shoe, I should say." In SIGN, we learn that he even wrote a monograph on the use of footprints for identifying criminals. However, footprints are just not as useful as, for example, fingerprints or DNA analysis. An episode of the television series "Hawaii Five-0" that first aired on April 18, 2011, made this point. Commander McGarrett collects a large plastic bag filled with shoes of suspects. Nothing comes of this effort to match a footprint at the crime scene. Today's mass-produced shoes are difficult to distinguish from each other. In Holmes's era, shoes were individually made and thus more distinctive (Wagner 2006, 142). This no doubt helped Holmes to put footprints to such good use for so many years. Now because of limited utility, footprints are nearly absent from modern books on crime-detection methods (Saferstein 1995; Moenssens et al. 1995).

Section 3.4 Handwritten Documents

You may not be aware that the deduction of a man's age from his writing is one which has been brought to considerable accuracy by experts.

—Sherlock Holmes, *The Reigate Squires*

INTRODUCTION

Alexander Cargill published an article entitled "Health and Handwriting" in 1890 (627–631). In it, he made claims that handwriting could be used to determine age, character, and perhaps even gender. He sent a copy of this article to Conan Doyle in December 1892. In June 1893, Conan Doyle published *The Reigate Squires* (REIG), the story in which Holmes makes his greatest use of handwriting. There, as we shall see, Holmes's deductions exceed what even Cargill claimed could be done with handwriting. In *The Red Circle* (REDC), Conan Doyle shows that he accepts Cargill's claim about gender and handwriting (see figure 3.2). Emilia, the hidden lodger, prints messages to the landlady out of concern that her gender would be revealed by her handwriting. She wants to conceal the fact that she has replaced the original renter, her husband Gennaro. Holmes uses his knowledge of handwriting analysis to deduce what is going on. Cargill's claim about being able to tell age from handwriting has been disputed (Rendall 1934, 79).

A number of actual cases have involved handwriting evidence, including one in which Conan Doyle himself played a major role. As we've seen, the reputation of Alphonse Bertillon suffered greatly when he used handwriting to pronounce that Dreyfus[5] had written the controversial memorandum in 1894 (Wagner 2006, 162). Here we look at a few well-known real cases before moving on to Holmes and handwriting.

REAL CASES

New York Zodiac Killer

In New York City in 1989, Heriberto Seda sent notes to the police announcing himself as the new Zodiac Killer. The original Zodiac Killer had claimed to have murdered thirty-seven people in San Francisco between 1966 and 1974. He had never been apprehended. Seda stated that he would kill one person from each of the twelve signs of the zodiac. His first note was sent on

[5] Numerous books have been written about this famous case.

Figure 3.2 Holmes and Watson discuss handwriting evidence

November 17, 1989, and his first attack occurred on March 8, 1990. He then set a pattern of attacking every twenty-one days or a multiple thereof. When that pattern became clear, police flooded the NYC boroughs of Brooklyn and Queens on July 12, 1990. Seda must have been alarmed by this, and no further attacks happened until August 1992. After just a few more attacks, the New York Zodiac Killer disappeared.

Then, on June 18, 1996, Seda shot his half-sister Gladys Reyes in the buttocks. She managed to get to a neighbor's apartment, and police were called. Seda was arrested and his numerous weapons collected. Seda wrote out a confession to the shooting of his half-sister. It was at this point that his handwriting was recognized by Detective Joseph Herbert: "[T]he t's curving to the left, the i's dotted to the right of the stem, the frequent underlining. I knew right away it was him."

Seda's fingerprints were also matched to those on several of the taunting zodiac notes he had sent to the police. Without the handwriting evidence, the police may never have even looked at the fingerprints. As further evidence, one of his zip guns was shown to have been the weapon that killed one of the victims. Seda was convicted of three murders and sentenced to eighty-three and a half years. Subsequently, in July 1999, he was given an additional 152.5 years for eight attempted murders. Without the use of handwriting, the New York Zodiac Killer may never have been apprehended.

Lindbergh Baby Kidnapping

When the O. J. Simpson case of the 1990s was called the "crime of the century," it was usurping the name of an earlier case: the Lindbergh baby kidnapping of March 1, 1932. A huge investigation followed the disappearance of Charles Lindbergh Jr. in New Jersey. A variety of evidence was used to convict Bruno Richard Hauptmann. Handwriting experts were involved. Footprints found under the nursery window were impossible to measure (some mud had been found in the nursery). There were no fingerprints.

When Hauptmann appealed his conviction, he was rejected because of three types of evidence: $13,760 of the ransom money was concealed in his garage (Behn 1994, 215)—serial numbers had been recorded on the $50,000 delivered; wood from his attic matched that in the homemade ladder used to reach the second-floor nursery; and there was handwriting analysis of the more than a dozen ransom notes sent to various people.

The handwriting experts were certain that the same person had written all of the notes because later notes referred to statements made in earlier ones. In addition, the second note was written on paper torn from the first note, and the torn edges matched. Similarities and misspellings were consistent throughout the notes. For example, money was spelled "mony," boat was "boad," and anything was "anyding." Finally, the handwriting was the same with "i's" rarely dotted and "t's" rarely crossed. All experts were unanimous that the same "German" had written all of the ransom notes. A strong statement on the handwriting evidence was given by Charles Appel at Hauptmann's grand jury hearing. Appel was the first full-time employee at the new FBI laboratory that had opened on November 24, 1932 (Fisher 1995, 9). He testified that he had examined 1,500 handwriting samples and never found any of the peculiarities that were present in both Hauptmann's handwriting sample and the ransom notes. Appel's conclusion was that it was "inconceivable that anyone but Hauptmann could have written the ransom notes" (Fisher 1995, 242). At Hauptmann's trial in 1935, Albert Osborn's testimony on the handwriting in the ransom notes "proved devastating for the defense" (Wagner 2006, 167). Osborn was the author of the most influential reference book on the identification of handwriting at the time.

Howard Hughes

In 1924, at age eighteen, Howard Hughes inherited a fortune that his father had made mainly by patenting a drill bit for the Texas oil fields. Hughes went on to even greater wealth as a movie producer and director beginning in 1927. He had a great interest in aviation and was a first-class pilot. He formed Hughes Aircraft in 1932 to build airplanes.[6] In 1939, Hughes acquired Trans World Airlines, which later merged with American Airlines. In 1948, he gained control of RKO Studios. In his later life, he ventured into the casino business in Las Vegas. When Howard Hughes died on April 5, 1976, his estate was estimated at two to three billion dollars (Freese 1986, 342). At the time of his death, Hughes had no wife, no children, no siblings, and no living parents. Soon a will emerged, but under suspicious circumstances with a suspicious provision. In this will, Melvin Dummar, a gas station owner from Gabbs, Nevada, was left one-sixteenth of Hughes's estate. This would amount to about $156,000,000 for someone never previously connected to Hughes. Needless to say, this will was examined for authenticity. Handwriting analysis by FBI Agent Jim Lile led to the conclusion that the will was a forgery (Fisher 1995, 250). One of Lile's points dealt with the flow of natural handwriting. The will contained a number of instances where the writing was interrupted. This is often the case in forgeries because of the need to refer to a writing sample being copied. Dummar was attempting to copy Hughes's writing as found in a handwritten memo and reproduced in the January 1970 issue of *Life* magazine. This memo contained thirteen capital letters of the alphabet. Dummar faithfully reproduced those letters in the will. But he used nine others, only two of which resembled Hughes's writing (Harris 1986, 375).

In addition to the handwriting evidence, Dummar was betrayed by fingerprints (Freese 1986, 347) and by being caught in a number of lies. He eventually confessed to writing the fake Howard Hughes will. This incident was the second time in the 1970s that someone had tried to forge the handwriting of Howard Hughes. Clifford Irving had previously tried to obtain a contract for an "autobiography." He managed to fool handwriting experts, but was found guilty of forgery on the basis of voiceprint evidence (Saferstein 1995, 474, 493).

Arthur Conan Doyle and George Edalji

Once his reputation as the author of the Sherlock Holmes stories became established, Conan Doyle was contacted for advice in actual cases (Stashower 1999, 255). Most often people asked him for help in finding a relative or a loved one who had disappeared (Booth 1997, 261). Sometimes jewels were involved (Booth 1997, 262). Once he even helped clear a dog accused of killing sheep (Stashower 1999, 255).

But Conan Doyle's most famous real-life case involved handwriting evidence. The involvement of Conan Doyle in the George Edalji case has been

[6] Recall the "Spruce Goose," now housed in Oregon.

fully described in several Conan Doyle biographies (Miller 2008, 257–272; Stashower 1999, 254–263; Carr 1949, 268–290). Here we take a brief look at Edalji's case, focusing on the handwriting evidence.

George Edalji's father had immigrated to England from Bombay, India (Miller 2008, 257). He became vicar of St. Mark's Church in Great Wyrley. In 1888, the Edalji family began receiving hate letters that were probably racially motivated (Miller 2008, 258). A family servant was accused, based on handwriting, of writing these letters. She confessed, and the letters stopped. In 1892, letters started arriving again, this time written by a different hand. In December 1895, the letters again stopped.

Between February and August 1903, sixteen animals were killed or mutilated in the Great Wyrley area, near Birmingham (Miller 2008, 261). Letters accusing George Edalji, a solicitor, of these crimes were received by the police. The Edalji family was again sent hate mail. Local authorities decided that George Edalji had written the very letters that identified him as the culprit. He also, they claimed, wrote the hate mail to his own family. On the basis of this fantastic conclusion, Edalji was arrested. He was accused of writing the letters and maiming a pony on the night of August 17, 1903. His trial began on October 20, 1903 (Carr 1949, 274). Despite having an alibi for the night of the pony maiming, and despite being present in the house when letters were slipped under the door of the Edalji home, George was convicted. "The police case rested on the testimony of a handwriting expert" (Stashower 1999, 256). Graphologist Thomas Gurrin had already helped convict an innocent man, Adolf Beck, in 1896 (Booth 1997, 264). George Edalji was sentenced to seven years of hard labor and was forced to spend his days breaking stones in a quarry (Booth 1997, 264).

The animal attacks continued for twenty-five years after Edalji was imprisoned (Miller 2008, 272). The authorities were unmoved. When this news spread, public interest resulted in the Home Office receiving a petition signed by 10,000 people protesting Edalji's conviction. With no appeal process in the legal system, a committee was appointed to look into the matter. They concluded that Edalji was not guilty of the animal attacks, but that he had written the letters (the ones accusing himself!). Edalji was not pardoned, but he was released from prison in October 1906. However, as a convicted felon, he was unable to practice law. Conan Doyle was not yet involved. When he did learn of what he considered to be an obvious injustice, he sprang into action. George Edalji had written a letter in November 1906, asking Conan Doyle for help (Miller 2008, 263). After reviewing all the evidence and meeting with George Edalji, Conan Doyle wrote articles and lectured to packed halls about his innocence (Coren 1995, 124). As a result of his efforts, Parliament asked the home secretary if a new inquiry would be made into the question of the handwriting in the anonymous letters (Miller 2008, 271). Soon Edalji was allowed to resume his work as a solicitor. George Edalji was one of the guests at Conan Doyle's marriage to Jean Leckie on September 18, 1907 (Stashower 1999, 260). In 1934, Enoch Knowles, "a labourer at an iron works," admitted to writing the letters. No one was ever arrested for killing and mutilating the animals.

Holmes's Usage

Handwriting issues arise in a number of the sixty stories. Several times Holmes uses the stylistic features of handwriting to date or identify it. For example, in HOUN, a mere glance enables him to correctly date the manuscript describing the Baskerville legend. He points out to Dr. Watson the alternating shape of the letter "s." In *A Case of Identity* (IDEN), Mr. Windibank is afraid that his writing will be recognized by his stepdaughter Mary Sutherland. So he corresponds with her by typed letters only. In *The Stockbroker's Clerk* (STOC), the Beddington brothers need to be able to duplicate Hall Pycroft's writing. So they get him to write a note accepting a position at the Franco-Midland Hardware Company in Birmingham. They then send him off to his new overpaid job, the fool's errand mentioned in section 1.2. Meanwhile one of them assumes his job in London, emulates his writing, and attempts an inside robbery there. In *The Valley of Fear* (VALL), Holmes receives a note that has the Greek "e" (epilson, ε) with a flourish at the top. Recognizing this feature of Porlock's hand, he knows it was written by his spy within Professor Moriarty's organization. Holmes then trusts the information about the danger to Mr. Douglas and knows that Moriarty is behind Douglas's subsequent death. In SIGN, Holmes verifies that all the notes Mary Morstan has received are from the same person. Thaddeus Sholto also uses the Greek epsilon and has twirls on the letter "s."

Ink blotters, so rarely used today, play a role in two of Holmes's cases. In TWIS, Holmes observes that the writer did not know the address to be put on an envelope. He cleverly deduces this by noting that the ink has dried without blotting in some places and not in others. The writer must have left the envelope to check on the address. Upon return, there was no need to blot as the ink had already dried. So we find in this case that the interrupted writing found in the Howard Hughes will was used over fifty years earlier by Holmes. In *The Missing Three-Quarter* (MISS), Holmes realizes that blotting paper has been used. He reads the writing backward and obtains more information on the disappearance of Godfrey Staunton, the rugby player who has gone missing.

There are three cases in which Conan Doyle appears to have been influenced by his reading of the article by Cargill, who claimed that gender and moral character could be discerned from handwriting analysis. In NAVA, he recognizes that the letter from Percy Phelps was written by a woman, and one of "rare character" at that. In CARD, he deduces that a man had addressed the package that contained the two severed ears.

Handwriting plays its largest role in REIG. Holmes makes amazing deductions from the torn corner fragment of the note clutched by the dead William Kirwin. Pushing well beyond what Cargill claimed (1890, 627–631), Holmes concludes that it was written by two persons, one of whom had a "strong" hand and was the leader. Additionally the strong person was younger, and the two people were related! Holmes bases this fantastic assertion of shared family on similarities in their writing, such as the use of the Greek epsilon. This goes well beyond his use of the Greek epsilon in SIGN and VALL, where he used it merely to recognize someone's writing.

at quarter to twelve
learn what
maybe

That Holmes's other conclusions have some merit can be seen by examining the above figure. He notes the "t's" in "at" and "to" are strong, while those in "quarter," "twelve," and "what" are weak. Holmes decides that the younger of the two authors had the strong hand and was the leader. Looking at the gaps

If you will only come round at quarter to twelve
to the east gate you will learn what
will very much surprise you and maybe
be of the greatest service to you and also
to Annie Morrison. But say nothing to anyone
upon the matter

between words, he sees that this man wrote his words first and left gaps for the others. This resulted in "quarter" being squeezed between "at" and "to." Using these conclusions, the Cunninghams, father and son, are deemed to be suspects. When the full note is recovered, it only serves to verify Holmes's conclusions.

Attempts have been made to identify the other twenty-three deductions Holmes makes about the writing in the note (Baring-Gould 1967, vol. 1, 343).

Finally, in NORW, Sherlock brilliantly deduces that Jonas Oldacre has written his will on the train coming in to London because most of the document is written in a very shaky hand, clearly during a bumpy train ride. But at two points, the writing is smooth. This happened when the train stopped at a station. In two other places, the writing is completely illegible when the train was crossing switches. Knowing that a serious will would never be written in this way, Holmes is immediately suspicious of Jonas Oldacre.

SUMMARY

Handwriting analysis has played a significant role in crime detection for many years. The FBI has had a documents department since its opening day in 1932 (Fisher 1995, 242). We've looked at cases ranging from the time of Conan Doyle up to the Zodiac case. Corporations use handwriting analysis during the hiring process. Analysts then attempt to determine the character, traits, and personality of the writer, with the hope that this process will lead to "good" hires. This is particularly true in Europe, especially in France (Rafaeli and Klimoski 1979). The use of handwriting to deduce things about people was pioneered by some monks in France in the 1880s (Edwards 1993, *The Sign of the Four,* 125). One of them, Abbe Jean Michon founded the French Society of Graphology in 1881. Despite the fact that graphology has been shown to be only 2 percent reliable (Schmidt and Hunter 1998, 265), approximately 80 percent of companies employ graphologists to analyze the writing in cover letters on job applications. These cover letters are required to be handwritten for this very purpose. Even the French government has been willing to pay a fee to graphologists, one of whom has stated that "the angle of the pen and the depth of the imprint can be used to detect the energy and libido" of the candidate.

There are stories of French job applicants trying to get around the handwriting analysis. An unemployed engineer who was having trouble getting an interview had his wife start writing the cover letters. He immediately started getting calls for jobs. In another case, forty-year-old Michel Malat had been rejected over 250 times, so he hired a graphologist to help him alter his handwriting (*The New York Times,* October 19, 1993, D1).

In the United States, those who claim to be able to discern personal qualities and character from handwriting are generally considered charlatans, because no scientific study has produced results establishing the validity of predicting personal traits from handwriting analysis (Blinkhorn 1993, 208). As one researcher stated, "We have tried our utmost but have failed to produce evidence to support the use of graphology for personal assessment." In a series of publications in *The Southern Literary Messenger* in 1836 and *Graham's Magazine* in 1841 (Sova 2001, 17, 48), Edgar Allan Poe poked fun at those who attempted to determine traits or character from handwriting. He presented fake handwriting samples of various people, such as Henry Wadsworth Longfellow, and then "analyzed" the person's traits. As recently as 2005, graphologists were embarrassed by their analysis of British Prime Minister Tony Blair (Wagner 2006, 167). They didn't realize that the writing sample was that of Microsoft founder Bill Gates.

Section 3.5 Printed Documents

> *It is a curious thing that a typewriter really has quite as much individuality as a man's handwriting. Unless they are quite new, no two of them write exactly alike.*
>
> —Sherlock Holmes, *A Case of Identity*

INTRODUCTION

Sherlock Holmes was a very early user of the idiosyncrasies of typewriters as an aid in his work. Even brand-new typewriters have unique features that allow forensic investigators to assign documents to a specific typewriter (Wagner 2006, 166). Wear and tear only increases the individuality of a typewriter. Letters can and do get bent, worn, and chipped.

What Mr. James Windibank did in IDEN may not have been a crime, but it surely was despicable behavior. Holmes's brilliant deductions that exposed Windibank still echo today in the cases of Alger Hiss and the Unabomber. The Alger Hiss case is yet another that was called the "trial of the century" (Hiss 1999, 15). This leaves us with at least three contenders for the most sensational crime of the twentieth century: the Lindbergh baby kidnapping, the Alger Hiss perjury case, and the O. J. Simpson killings. This century's title may already belong to the destruction of the towers at the World Trade Center and deaths of nearly 3,000 people on September 11, 2001.

REAL CASES

Alger Hiss

The first famous case involving typewriter evidence was that of Alger Hiss (1950). So much has been written about the case that we shall only briefly describe it, emphasizing those aspects that deal with his typewriter. Hiss graduated from Harvard Law School, where the famous Felix Frankfurter was his mentor. He then took a job as secretary to Supreme Court Justice Oliver Wendell Holmes (Jacoby 2009, 46). With such an outstanding pedigree, Alger Hiss was soon working for the U.S. government and advancing up the ladder of success. He moved from the Department of Agriculture to the Justice Department. Soon Hiss took a position in the office of the assistant secretary of state. Some have claimed that he was willing to move to the State Department and take a 25 percent salary cut in order to have access to materials that were of interest to the Soviet Union (White 2004, 41). Hiss began copying classified documents and passing them on to his contact, Whittaker Chambers. Some of these documents were handwritten, and others were typed.

In 1948, Chambers, having left the Communist Party that he had joined in 1925 (Jacoby 2009, 41), testified before the House Committee on Un-American Activities. He stated that Alger Hiss had been a Communist agent in the 1930s. As evidence, Chambers produced copies of some of the documents that Hiss had given him for delivery to the Soviets. Hiss's handwriting was obvious, and the typewritten documents matched a typewriter that the Hiss family had owned. Documents experts, finding a distinctive "e" and "g" (Wagner 2006, 168), were able to sufficiently prove that classified documents had been typed on this typewriter. That the typewriter belonged to Hiss was established by comparison with correspondence from him to an insurance company and a school (White 2004, 65). Experts for both prosecution and defense stated that the typewriter was the "most sensational" piece of evidence against Alger Hiss (White 2004, 70) (see figure 3.3). Hiss himself always considered the typewriter to be the biggest piece of evidence against him (Jacoby 2009, 140).

In 1950, Hiss was convicted of two counts of perjury for lying to the grand jury. A charge of espionage was not possible because the statute of limitations had run out on that offense. Alger Hiss spent forty-four months in jail (Hiss 1999, 6). After his release, he was able to regain his license to practice law. He spent the rest of his life trying to disprove the charges against him. In a 1978 motion, his lawyers suggested that the FBI had built a typewriter to match the one on which the State Department documents had been typed (White 2004, xv). One book on the FBI lab claims that it "proved impossible" to duplicate the Hiss typewriter (Fisher 1995, 266). It was even claimed that the defense was able to build a typewriter that could not be distinguished from Hiss's typewriter (Koppenhover 2007, 50). The obvious implication is that typewriters are not necessarily unique. Following his death in 1996, new information seemed to suggest that Hiss was indeed an agent for the Soviet Union (White 2004, xvii).

The Woodstock typewriter used as an exhibit at both of Hiss's perjury trials. It was first produced by the Hiss defense.

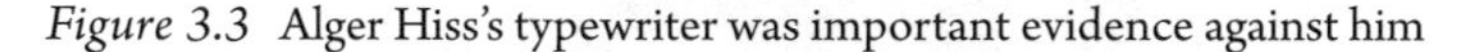

Figure 3.3 Alger Hiss's typewriter was important evidence against him

Unabomber

Ted Kaczynski was born in 1942. His IQ was higher than 160 (Douglas and Olshaker 1996, 82). He was given a scholarship to Harvard and graduated in three years. He next earned an M.S. and a Ph.D. in mathematics from the University of Michigan. He accepted an appointment as assistant professor at the University of California, Berkeley. But after this rapid beginning, Kaczynski abruptly left the academic world in 1969 (Douglas and Olshaker 1996, 90). In 1971, he moved into a remote area of Montana. There he eventually built a one-room shack, 10 feet by 12 feet. Having no electricity, he used a typewriter to write his messages to newspapers when he started mailing bombs.

Beginning in 1978, the Unabomber (whose name came from the FBI's code for the case—UNABOM—before Kaczynski's identity was established) killed three people and injured twenty-three others. His initial targets were university professors and airlines (Fisher 1995, 110). In 1995, he sent his "manifesto" to *The New York Times,* which published it on June 28, 1995. Kaczynski's brother David recognized some of the phrasing from letters that Ted had written to their mother. Eventually David notified the FBI. When they arrested the Unabomber at his Montana shack, they found three typewriters. One of them matched the idiosyncrasies found in his letters to the newspapers. This evidence was considered to be some of the most useful in the case that led to his conviction.

HOLMES'S USAGE

Only in IDEN does Sherlock Holmes utilize the idiosyncrasies of typewriters in his work. The bizarre story of Mary Sutherland sparked his interest with its

Figure 3.4 Somehow Mary Sutherland didn't realize that the man dancing with her was her stepfather, who lived with her!

uniqueness. Miss Sutherland had a nice inheritance from an uncle. But she lived on what she earned typing and gave the inheritance money to her mother and stepfather, Mr. Windibank. Needless to say, Mr. Windibank was anxious to keep Mary unmarried and thereby retain use of her money. So when she begins to show some interest in socializing, he devises the scheme that sends her to consult Holmes.

When Mary Sutherland insists upon attending the gasfitter's ball, Windibank also attends disguised as Hosmer Angel (see figure 3.4). There he courts his own stepdaughter. Afraid that Mary will recognize his writing, Windibank continues the relationship by means of typewritten letters. After making her fall in love with him, Windibank has Angel disappear. The heartbroken Miss Sutherland consults Holmes. The circumstances she describes, including the fact that Hosmer Angel's letters even have typed signatures, make Holmes suspect Windibank immediately. He corresponds with

Windibank, seeking an appointment. When he receives a typed response, Holmes has his case. From the idiosyncrasies of his typewriter, Holmes determines the truth. He gives Windibank the sternest of warnings, but has no legal cause for arrest.

SUMMARY

The usefulness of typewriter comparisons as evidence was not yet established when Conan Doyle published IDEN in 1891 (Wagner 2006, 166). FBI typewriter analysis began in 1933 and immediately led to the capture of a woman who had mailed poisoned fudge to a veterans hospital (Fisher 1995, 266). Foreshadowing the Unabomber was the 1989–1990 VANPAC case (Fisher 1995, 4). In this case, homemade bombs were mailed to the intended victims. The name of the case was derived from the fact that a Judge *Van*ce had been mailed an explosive *pac*kage. Again a typewriter flaw found by the documents department of the FBI was the initial discovery on the way to the arrest of Walter Leroy Moody. Likewise, the conviction of Alger Hiss was due largely to typewriter irregularities (Fisher 1995, 266). If the well-read Hiss had only read Holmes's quote about the individuality of typewriters, he may have been more careful in his espionage.

Even though the typewriter has been largely replaced by the computer, laser printer, and photocopier, documents still have distinctive features. Laser printer drums often have imperfections that create telltale marks on documents (Hudson 1994). The same applies to photocopiers. One FBI agent warned: "People believe photocopies are not traceable, and that's just not true" (Fisher 1995, 267).

Section 3.6 Cryptology

One if by land and two if by sea

—Henry Wadsworth Longfellow, *Paul Revere's Ride, 1861*

INTRODUCTION

Edgar Allan Poe had a lifelong interest in cryptograms. In *The Gold Bug* (GBUG), a story for which Poe won a $100 prize in 1843 (Silverman 1991, 205), he made the solution of a cipher the key plot element. Conan Doyle does the same thing in *The Dancing Men* (DANC). Both authors employ a substitution code and have their protagonist use frequency analysis to solve the problem.

Even prior to GBUG, Poe had published cryptograms as a challenge to his readers. The first of his several cryptology articles appears in the December 1839 issue of *Alexander's Weekly Messenger* (Sova 2001, 61). Then in 1842, in *Graham's Magazine,* Poe published two coded texts to further challenge his readers. He never published the answers, and they were not solved for 150 years. The ciphers published by Poe in 1842 were not simple. The first and easier one was not solved until 1992. It had probably been forgotten until a graduate student and Poe expert at Duke University named Terence Whalen solved it. The cryptogram had the message backward. Each letter corresponded to just one letter of the alphabet. The answer turned out to be a quote from a 1713 play, *Cato,* by Joseph Addison.

The more difficult second Poe cipher required several more years and a computer before giving up its secret. In 1996, Shawn Rosenheim, a Poe scholar from Williams College, established a prize for the person who solved it. The prize of $2,500 was won in 2000 by Gil Broza, a computer programmer from Toronto, Canada.[7] In this complicated cryptogram, the letter "e," for example, has fourteen different meanings; "z" has two different meanings. Broza started with the hypothesis that three-letter words were either "the," "and," or "not." This enabled him to get four of the letters of the word "afternoon," which he then guessed. Proceeding in this manner and with the speed of the computer, he gradually unraveled the rest of the text.

REAL CASES

Codes have always been used in wars. Conan Doyle employs two such codes, both from the American military. In the Revolutionary War (1775–1783), the American General Benedict Arnold made himself the everlasting symbol of perfidy to Americans by passing information to the British. He would send coded messages to a Tory[8] friend in Philadelphia. The simple code employed had been used for many years. Both users had a copy of the famous law book, *Blackstone's Commentaries on the Law of England.* The messages consisted of a series of numbers that led to words in that book. Each word was described by three numbers: page, line, and word number in that line (Butler and Keeney 2001, 68). Such codes are very difficult to break as the book being used is generally unknown to anyone but the users.

Here is an actual example of a Benedict Arnold message dated July 12, 1780:

> *120.9.7, W---- 105.9.5's on the.22.9.14.-- /of 163.8.19 F-- -172.8.7s to 56.9.8 |30.000| 172.8.70 to 11.94. in / 62.8.20. If 179.8.25, 84.8.9'd, 177.9.28. N-- is 111.9.27.'d on / 23.8.10. the 111.9.13, 180.9.19 if his 180.8.21 an.179.8.25., 255.8.17. for / that, 180.9.19, 44.8.9—a—is the 234.8.14 of 189.8.17. I--/44.8.9, 145.8.17, 294.9.12, in 266.8.17 as well*

[7] *The New Yorker,* November 27, 2000, 38.

[8] An American sympathetic to the British cause.

> *as, 103.8.11, 184.9.15.-- /80.4.20.-- I149.8.7, 10.8.22'd the 57.9.71 at 288.9.9, 198.9.26, as, a / 100.4.18 in 189.8.19—I can 221.8.6 the 173.8.19, 102.8.26, 236.8.21's - -/and 289.8.17 will be in 175.9.7, 87.8.7 - -the 166.8.11, of the.191.9.16 / are.129.19.21 'of - -266.9.14 of the.286.8.20, and 291.8.27 to be an - -163.9.4 / 115.8.16 -'a.114.8.25ing - -263.9.14. are 207.8.17ed, 125.8.15, 103.8.60 - -/from this 294.8.50, 104.9.26—If 84.8.9ed—294.9.12, 129.8.7. only / to 193.8.3 and the 64.9.5, 290.9.20, 245.8.3 be at an, 99.8.14 . / the.204.8.2, 253.8.7s are 159.8.10 the 187.8.11 of a 94.9.9ing / 164.8.24, 279.8.16, but of a.238.8.25, 93.9.28.*

The decoded result:

> *General W[ashington] - -expects on the arrival of the F[rench] - -Troops to collect / 30,000 Troops to act in conjunction; if not disappointed, N[ew]. York is fixed / on as the first Object, if his numbers are not sufficient for that Object, / Can-a- is the second; of which I can inform you in time, as well as of / every other design. I have accepted the command at W[est]. P[oint]. As a Post in which / I can render the most essential Services, and which will be in my disposal. / The mass of the People are heartily tired of the War, and wish to be on / their former footing—They are promised great events from this / year's exertion—If—disappointed—you have only to persevere / and the contest will soon be at an end. The present Struggles are / like the pangs of a dying man, violent but of a short duration—*

In 1780, Arnold's British contact, a Major Andre, was captured, and papers exposing Benedict Arnold were found. Arnold fled and spent two years fighting with the British Army. He eventually retired to London and died there a few years later.

During the Civil War, a Union sympathizer, J. O. Kerbey, managed to gain employment with the Confederacy as a telegrapher. He proceeded to use a courier to send coded messages containing military information to Washington. In his code, every fifth word was to be read (Butler and Keeney 2001, 120). This same type of code was the basis for the "Captain Midnight" decoder rings that were popular with young viewers of the radio show in the 1940s. Each week, young members of the Secret Squadron would receive a message that would reveal hints about the plot of the upcoming episode. They would be told to use their decoder rings and then read, for example, every tenth word in order to decode the message.

Another military coding strategy was the World War II use by the U.S. military of the Navajo code talkers. This ingenious code merely used Navajo, a spoken and not written language. Almost no one outside the tribe knew the language. About 200 young Navajo men were employed in this effort.

During training, a group of thirty-two men developed words for military things that were not even in their regular language. It is thought that not even Navajo speakers could have understood them (Butler and Keeney 2001, 77). The Navajo code was declassified in 1968 (Nez 2011). The code had a few hundred Navajo words for common terms and spelled out any other words needed. For example, the letter "a" was represented by any of three Navajo words whose English meaning was a word starting with "a" (Nez 2011, 103). The Navajo code talkers were active in the Pacific theater. They sent messages about troop locations, calls for ammunition, food, medical supplies, and any other information deemed important enough for coding. In at least one operation, the command declared that only Navajo code would be used (Nez 2011, 189).

An interesting case of a code resulting in a criminal conviction is that of Patty Hearst in 1974. She was kidnapped by the so-called Symbionese Liberation Army (SLA). Later she was photographed participating in a bank heist with SLA members. Coded messages were used by this group. Hearst was convicted of armed robbery because she had been given access to the SLA code. This was taken as proof that they trusted her. Therefore her participation was considered to be voluntary (Fisher 1995, 272). Her original sentence of thirty-five years was commuted, and she was released in February 1979, having served twenty-two months. In 2001, she received a full pardon from President Bill Clinton.

HOLMES'S USAGE

Conan Doyle first uses a cipher in a Holmes case in his nineteenth effort, *The "Gloria Scott"* (GLOR). This is also Holmes's first code to crack, as he tells us that this is the first case he solved while he was a college student.[9] Victor Trevor, the only friend Holmes made during his two years in college, invites Holmes to spend the long vacation between terms at the Trevor estate. Victor's father, Old Trevor, is so amazed by what Holmes deduces about him that he tells Holmes:

> I don't know how you manage this Mr. Holmes, but it seems to me that all the detectives of fact and of fancy would be children in your hands. That's your line of life sir.

[9] Sherlockians have amused themselves by trying to construct a chronology of Holmes's cases. This is made difficult, and perhaps more enjoyable, by the fact that Conan Doyle was careless in this regard.

Soon things go bad for Old Trevor. A colleague from the past, Hudson, turns up and blackmails him, staying for weeks. After Victor drives Hudson away, a coded note from Old Trevor's friend Beddoes soon arrives:

> The supply of game for London is going steadily up. Head-keeper Hudson, we believe has been now told to receive all orders for fly-paper and for preservation of your hen-pheasant's life.

After reading the coded note, Old Trevor, a.k.a. James Armitage, has a stroke and dies.[10] Holmes quickly solves the cipher, realizing that the code merely consists of instructions to read only every third word, à la Captain Midnight. This converts the nonsensical wording in the message to "The game is up. Hudson has told all. Fly for your life."

Armitage had been convicted of embezzling from the bank where he was employed in London. Sent to Australia aboard the *Gloria Scott,* he escaped by participating in a mutiny. Now having his freedom, he changed his name to Trevor, returned to England, and prospered. Hudson was the lone seaman who had survived when the *Gloria Scott* sunk at sea. He eventually found Trevor and extracted money and support in return for his silence about Trevor's past as James Armitage.

In VALL, Holmes is getting information from within the Moriarty organization. He receives a coded message in Fred Porlock's handwriting:

> 534 C2 13 127 36 31 4 17 21 41 Douglas 109 293 5 37 Birlstone 26 Birlstone 9 127 171

Holmes deduces that Porlock is using page 534 and column two of a book (see figure 3.5). Unlike Benedict Arnold, Holmes does not know which book is being used. But he is able to brilliantly deduce the title. The book must be common enough that Porlock would be sure Holmes had a copy. It must be large enough to have at least 534 pages and two columns. The numbers then tell the words. Since the words "Douglas" and "Birlstone" do not appear on page 534 in the book, they are written out. Using an old version of *Whitaker's Almanac,* Holmes soon learns that Douglas is in danger from Professor Moriarty:

> There is danger may come very soon one Douglas rich country now at Birlstone House Birlstone confidence is pressing

Alas it is likely that it wasn't only Douglas who was in danger. We never hear of Porlock again.

[10] Perhaps Armitage's death is another example of the Baskerville Effect?

Figure 3.5 Sherlock Holmes was able to deduce what book Porlock had used in his cipher message in *The Valley of Fear*.

The cipher used in VALL is a type known as an Arnold cipher, or sometimes an Abner Doubleday cipher. Doubleday, a Union general in the American Civil War, is considered by many to be the inventor of baseball. A recent biography of Abner Doubleday (Barthel 2010, 1) denies that he had anything to do with baseball. He did, however, use a cipher system very similar to the one employed by Benedict Arnold.

Abner Doubleday was stationed at Fort Sumter just before the first shot in the American Civil War was fired there. He and his brother Ulysses, a New York banker, feared that their correspondence was being intercepted and read. So Abner proposed that they code their messages using the exact same edition of a dictionary (Barthel 2010, 57). Then three numbers would define a word: page number, column number, and word number from the top. The Doubledays used this code from September 1860 to March 1861. The use of a dictionary and multiple pages pretty much allowed the brothers to find exactly the words they wanted. In VALL, Porlock is hampered by being restricted to a single page of the almanac. Thus the phrase "confidence is pressing" is as close as Porlock could come to conveying a sense of urgency in his message.

Abner Doubleday described their cipher in his 1876 book titled *Reminiscences of Forts Sumter and Moultrie* (Klinger 2006, vol. III, 637). Was the well-read Conan Doyle aware of the Arnold cipher and Doubleday's book? Conan Doyle does mention the American Civil War in three Holmes stories. But if he was aware, why didn't he use the more flexible approach of allowing multiple pages to be cited?

In REDC, Conan Doyle employs a code almost as simple as Paul Revere's. The Red Circle is a secret Italian political terror organization (Bunson 1994, 208). Gennaro, regretting that his membership in the Red Circle has put himself and his wife Emilia in danger, has hidden her in Mrs. Warren's boarding house. The landlady is alarmed by the fact that her new tenant never leaves the room, even for meals. So she consults Holmes. Gennaro communicates with Emilia by means of the agony column[11] in the *Daily Gazette*.[12] Holmes is an ardent reader of the agony column. He uses it in several cases, and in *The Noble Bachelor* (NOBL), he even remarks that all he reads in the newspaper is the criminal news and the agony column. Soon Holmes is reading Gennaro's messages to Emilia. In one of his newspaper messages, Gennaro gives Emilia the code: One flash of light means A, two means B, etc. He tells her to look for candle flashes that night from the third-floor window of the tall red house across the street from Mrs. Warren's boarding house. Holmes is there to intercept the message and soon solves the case. In fact, at the conclusion Holmes sends Emilia a candle message to come over to the red house.

Here is one of many instances where Conan Doyle is less than careful with details. Because the Italian alphabet has no letter "k," the signal sent by Gennaro does not actually spell the warning "attenta" (beware). For example, twenty flashes do not indicate "t," but rather "u." The result is a nonsensical message. But Sherlockian scholars are always ready with an ingenious explanation. One suggestion (Yates, D. A., in Klinger and King 2011, 295) is that Emilia and Gennaro, in order to confuse the Red Circle (but not Sherlock Holmes!), agree to use the English alphabet to spell Italian words.

Conan Doyle's most extensive use of cryptology occurs in DANC, where Poe's strong influence on Conan Doyle is again evident. In fact, Conan Doyle stated that "all cryptogram solving yarns trace back to" Poe's GBUG (Fowler 1994, 363). Both authors use a substitution cipher that is solved using frequency analysis. The two ciphers provide another example of the relative clarity of Conan Doyle's writing. Poe's presentation of the code in GBUG is confusing to the reader. Here it is:

> 53‡‡†305))6*;4826)4‡.)4‡);806*;48†8¶60))85;1‡(;:‡*8†83(88)5*†
> ;46(;88*96*?;8)*‡(;485);5*†2:*‡(;4956*2(5*—4)8¶8*;4069285);)6
> †8)4‡‡;1(‡9;48081;8:8‡1;48†85;4)485†528806*81(‡9;48;(88;4(‡?3
> 4;48)4‡;161;:188;‡?;

[11] Today's personal ads.

[12] REDC is one of fourteen Holmes stories in which advertising in newspapers is mentioned.

When decoded the message contained directions telling where on Sullivan's Island to dig for a hidden treasure. William Legrand followed the instructions and successfully recovered a great fortune in gold and jewels.

In GBUG, Legrand assumed that the most common character in the coded message, the number 8, represented the most common letter, "e." He then found no less than seven instances in the message where the three-character sequence ";48" appeared. Assuming this to be the word "the," Legrand was well on his way to the solution. He was then able to find the hidden treasure.[13]

Holmes uses a very similar approach. Initially he is hampered by lack of data. In GBUG, the single coded message contained 193 characters. Initially Holmes has only the fifteen characters of the first message from Chicago gangster Abe Slaney to his former fiancée Elsie Cubitt: "Am here Abe Slaney."

Holmes applies his knowledge of statistics and waits for more data. Finally after five messages totaling sixty-two characters, he solves the cipher. Like Legrand, Holmes also assigns the most frequent of the sixty-two characters to the letter "e." There are seventeen "e's" in the sixty-two letters in the five messages. He deduces that a man holding a flag is the last letter in a word. Then he notices that the fourth message is just a five-letter word with "e" in both the second and fourth positions:

NEVER

He considers the word "never" to be a more likely choice than "sever" or "lever."[14] Next Holmes realizes that Elsie's name might very well be included in a message.

[13] In *The Musgrave Ritual*, Holmes also solves a cryptic message and recovers a treasure (Hodgson 1994, 213).

[14] On the other hand, Conan Doyle has Holmes ignore some other words that have "e" in the second and fourth positions, including such promising ones as "seven" and "jewel."

When the fifth message contains a five-letter word beginning and ending with "e," Holmes already has the "l," "s," and "i":

Soon Holmes has the solution.[15] But the fifth message sounds ominous,

"Elsie, prepare to meet thy god."

Holmes rushes to the Cubitt home but is too late to prevent the death of Hilton Cubitt. Although Holmes solved the cipher, this case can hardly be considered a success. Abe Slaney kills Hilton Cubitt. Elsie Cubitt, in despair over her husband's death, shoots herself in an unsuccessful attempt at suicide. Holmes's apprehension of Abe Slaney is somewhat anticlimactic. But there is some poetic justice in that Holmes lures Abe Slaney to come to him by leaving a message in Slaney's own code, where the flags held by the stick men signal the end of a word:

COME HERE AT ONCE

Slaney, thinking the message is from his beloved Elsie, walks right into the trap and is captured.

Those who lament Conan Doyle's lack of attention to detail should take note that he has Holmes state the order of the frequency of letter usage to be "E, T, A, O, I, N, S, H, R, D, L." This exactly matches the true list of the first eleven letters. Poe, however, is way off. His list in GBUG is "E, A, O, I, D, H, N, R, S, T, U" (Fowler, in Hodgson 1994, 358).

[15] It was noticed long ago that the dancing men used for the letter "V" in message four and those used for the letter "P" in message five are identical. Recently it has been determined that the print shop faithfully reproduced what they were given, so this error can now be attributed to Conan Doyle. Of course, Sherlockians would say that Watson wrote it down wrong (Klinger 2011, 24).

SUMMARY

Just as with footprints, Conan Doyle uses a diverse array of ciphers in the Holmes stories. As we have seen, the codes he used are relatively simple ones. Holmes however, is certainly capable of solving more complicated ciphers. He informs us in DANC that he has written a monograph in which he analyzed 160 different ciphers.

Section 3.7 Dogs

Dogs never make mistakes.

—Sherlock Holmes, *Shoscombe Old Place*

INTRODUCTION

The most famous dog to appear in the Holmes stories is, of course, the hound of the Baskervilles. In 1742, Hugo Baskerville had his throat torn out by a dog "larger than any hound that ever mortal eye has rested upon." The family is then plagued by the hound of the Baskervilles for more than 150 years. This unrealistic time-span makes attentive readers question whether there is a dog in the story at all. But at the end of chapter 2, we're told of the footprints of a "gigantic hound." In Holmes's time, Sir Charles Baskerville is chased by the hound until he dies of fright. A number of other dogs appear in the Holmes stories. Several of them play a role in Holmes's work: Toby in SIGN, Pompey in MISS, Roy in *The Creeping Man* (CREE), Carlo the poisoned spaniel in *The Sussex Vampire* (SUSS), Mrs. Hudson's terrier in STUD, Lady Beatrice Falder's spaniel in *Shoscombe Old Place* (SHOS), and the unnamed hound in *Silver Blaze* (SILV). In CREE, we learn that Holmes considered writing a monograph on the use of dogs in detective work.

WORKING DOGS

Dogs have been used in forensic investigations as long ago as 400 B.C. (Gerritsen and Haak 2007, 20). After some eighteenth-century police dog activity in Belgium, their use was banned in 1793. More than a century passed before the use of dogs for police work resumed (Gerritsen and Haak 2007, 23). Police dogs generated a lot of favorable publicity beginning with an identification of a murderer in Germany in 1903. There followed a number of other successful cases in the next decade (Gerritsen and Haak 2007, 26). A good candidate for the most famous police dog is Rex III who, in 1950s England, made more than 130 "arrests"

(Lane 2005, 55). Originally German shepherds were the breed of choice for police work. In fact, the Royal Canadian Mounted Police Dog Section, formed in 1935, still uses only German shepherds (Burhoe 2007). The first police dog for the Canadian Mounties was named Dale. Upon being put into service, he immediately tracked and caught a car thief. In 1910, the Rottweiler was named the official breed for police work in Germany. In recent times, the Malinois has been preferred (Gerritsen and Haak 2007, 117). Since 1960, identification by a police dog has been taken as acceptable proof in Scotland (Puttnam 1991, 24). The Dutch Supreme Court has made a similar ruling.

In the United States, the use of police dogs is governed by the individual states. For example, Connecticut started its canine corps in 1937 (*Police Chief*, January 1991, 50), while Virginia began its in 1961 (*Police Chief*, October 1991, 60). In recent years, police dogs have provided assistance in locating drugs. Their sense of smell is a thousand times more sensitive than humans' (PBS, "Dogs That Changed the World," *Nature*, October 2011). Their sniffing ability has been employed by U.S. Customs where beagles have been used. Police in Ohio have even used Chihuahuas to detect marijuana (Jackson 2009). The evolution of police dog responsibilities in Connecticut has been typical. Initially the dogs were used for criminal detection and perhaps crowd control. In 1967, the dogs' duties were expanded to include narcotics detection. Starting in 1971, detection of explosives was added to their expertise. In 1979, the dogs started doing searches for bodies. Finally, in 1986, they began detecting accelerants used in arson, where they exceeded the success rate of the mechanical devices they replaced (*Police Chief*, October 1991, 60–65).

New York City has two kinds of police dogs. Patrol dogs are generally German shepherds or Malinois; detection dogs are usually Labrador retrievers. A study done in Michigan in 2000 showed that the success rate of detection dogs was 93 percent. They far outperformed the teams of two to four police officers, who managed only a 59 percent rate (Bilger 2012, 55). The dogs were also five to ten times faster than the humans.

Working dogs have frequently been used by the military. In World War I, Germany used over 30,000 dogs (Bilger 2012, 48). The most famous dog helping the American soldiers in World War I was Stubby. He served mainly in France, where his acute hearing and sense of smell frequently aided the soldiers. In 2006, Stubby was again honored, this time with a brick in the Walk of Fame at the World War I museum in Kansas City. Stubby's World War II counterpart was Chips. He also served in Europe, where he reportedly captured enemy soldiers on his own (America Comes Alive website). In World War II, over 10,000 dogs, most of them Doberman pinschers, were employed by the U.S. military in the Pacific theater. There, no camp guarded by a dog was ever subject to a surprise attack (Gerritsen and Haak 2007, 196). Very recently, a Belgian Malinois named

Cairo accompanied the U.S. military mission into Pakistan that resulted in the death of the terrorist Osama bin Laden (Schmidle 2011, 35). Cairo's assignment was twofold: He would help deter any curious neighbors around the perimeter of the bin Laden compound. Secondly, he would search inside for false walls and hidden doors should that prove necessary. Cairo participated in all the training drills in North Carolina and Nevada (Schmidle 2011, 39). Cairo's work earned him an introduction to the president of the United States. The American military services now have about 3,000 active-duty dogs (Bilger 2012, 48).

HOLMES'S USAGE

In SIGN, Holmes sends Watson to fetch Toby, a "half spaniel and half lurcher." Given the task of tracking Tonga, who has stepped in the odiferous creosote, Toby soon shows that he is no bloodhound. Instead of Tonga, Toby leads Watson to the creosote factory.[16] But Holmes's faith in dogs never wavers. In MISS, Holmes tries again. This time he uses Pompey, who is part foxhound and part beagle.[17] Pompey's job is to find the missing Godfrey Staunton (see figure 3.6). Holmes has used a syringe to squirt aniseed oil onto the back wheel of Dr. Armstrong's carriage. Pompey is up to the task and follows the aniseed odor directly to the cottage where the distraught Staunton is found with his recently deceased wife. Perhaps aniseed is easier to track than creosote.

The Creeping Man (CREE) is the story in which Holmes and Watson discuss the use of dogs in detective work. Holmes tells Watson that he is considering writing a monograph on the subject. Professor Presbury is the creeping man, and when his wolfhound Roy[18] attempts to bite him on several occasions, he consults Holmes, who bases his conclusions partly on the dog's ability to detect significant changes in his master. Courting a woman much younger than he, the professor is taking a serum that confers great strength and agility, but alters his personality (and perhaps his odor) as well. His taunting of Roy nearly cost the professor his life when the chained dog got loose and in a rage almost killed Presbury. A similar event occurs in *The Copper Beeches* (COPP) when the starving mastiff Carlo attacks and nearly kills his owner, Jephro Rucastle.

Another case of a dog detecting a change in his master occurs in the sixtieth and last story, SHOS. The role of the dog is of such importance that the story was originally called *The Adventure of the Black Spaniel* (Holroyd 1959, 49). Here a spaniel belonging to Lady Beatrice Falder is lonely for its

[16] Toby does a better job tracking Professor Moriarty in Nicholas Meyer's *The Seven- Per-Cent Solution*.

[17] Perhaps Toby was fired?

[18] Conan Doyle himself had a dog named Roy.

Figure 3.6 Pompey successfully follows the trail of aniseed.

mistress. Holmes, suspecting that Lady Beatrice is dead, uses her spaniel to verify that the veiled passenger in the carriage is someone else. First elated at the sight and perhaps the odors from the approaching carriage, the dog snarls upon getting close and realizing that Lady Beatrice is not inside. In this way, Holmes verifies his hypothesis and the solution to the case soon follows.

In SUSS, there is the reverse instance of Holmes detecting a change in a dog. The other Carlo in the Canon belongs to Robert Ferguson. Holmes deduces instantly upon seeing Carlo drag his rear legs that the spaniel has been used as a test case for the administration of poison. This poison is discussed in chapter 5.

In LION, the death of Fitzroy McPherson's dog is the clue that leads Holmes to realize the true meaning of the phrase "Lion's Mane." Why did the Airedale die in the same manner and at the same place as his master? And when Ian Murdock nearly meets a similar fate at that same dangerous spot, Holmes belatedly recalls the dangerous and poisonous *Cyanea capillata*. Slow to act, Holmes nearly fails in this case. But eventually he solves the crime and exonerates all humans from blame.

In STUD, another dog, Mrs. Hudson's terrier, dies. Holmes, in accordance with the landlady's wishes, administers a poison pill that kills the sickly dog. In doing so, he demonstrates the method by which Jefferson Hope murdered Enoch Drebber. Hope let Drebber choose one of two pills and then took the other pill himself. When it was Drebber who died, Hope's beloved Lucy Ferrier was avenged. In a similar way, Carlo in SUSS is administered poison. The unbalanced, deformed teenager Jacky is filled with hate for his healthy new half-brother. He wants to administer curare to kill the toddler and uses Carlo as a test case to help figure out the needed dosage of the poison. Carlo survives. Holmes is able to unravel the facts and demonstrate that Mrs. Ferguson is in fact a loving mother and innocent of the attack on the baby.

Finally, in SILV, Holmes immediately realizes the importance of the silence of the hound during the night. Scotland Yard's Inspector Gregory is baffled by the facts of the case. His request for help from Holmes gives rise to this famous exchange:

> INSPECTOR GREGORY: Is there any point to which you would wish to draw my attention?
> SHERLOCK HOLMES: To the curious incident of the dog in the night-time.
> INSPECTOR GREGORY: The dog did nothing in the night-time.
> SHERLOCK HOLMES: That was the curious incident.

Later Holmes reveals that the horse quietly let itself be taken out on the moor in the middle of the night because it was John Straker himself, Silver Blaze's trainer, who led him there. On the moor, Straker tried to snip his own horse's tendon. That was when Silver Blaze rose up and struck Straker on the head with his hoof, instantly killing him. Why would a trainer attempt to injure his own horse? Straker had bet heavily on the opposition horse. He did this because he was in need of money to spend on his rather expensive mistress.

SUMMARY

Conan Doyle offers the reader an interesting array of dogs. In SIGN, the dog fails to track the person. In MISS, the dog succeeds in finding the person. In CREE, the dog detects a major change in his master. In SHOS, the dog demonstrates the absence of the person. Two dogs, in CREE and COPP, attack their masters, but for different reasons. Two other dogs are poisoned in SUSS and STUD, both helping Holmes solve the cases. Perhaps the most interesting dog of all is the nameless hound in SILV. He did nothing in the nighttime, and thereby gave rise to the most famous lines in the entire Sherlock Holmes Canon. Holmes, a believer in dogs right to the end, claims in SHOS that dogs *never make mistakes*. Apparently he has forgotten Toby's failure, so long ago, in SIGN.

It is notable that once again Holmes is a pioneer in a forensic technique. As we have seen, canine corps in police departments are a rather recent innovation, with most created in the twentieth century, long after Holmes used Toby.

Section 3.8 Conclusion

When it came to forensic science, Arthur Conan Doyle was an innovative thinker. That his contemporaries believed so is evident from this quote published in the *Illustrated London News* on February 27, 1932, nineteen months after Conan Doyle's death:

> Many of the methods invented by Conan Doyle are today in use in the scientific laboratories. Sherlock Holmes made the study of tobacco ashes his hobby. It was a new idea, and now every laboratory has a complete set of tables giving the appearance and composition of the various ashes.
>
> Mud and soil from various districts are also classified much after the manner that Holmes described.
>
> Poisons, handwriting, stains, dust, footprints, traces of wheels, the shape and position of wounds, the theory of cryptograms—all these and other excellent methods which germinated in Conan Doyle's fertile imagination are now part and parcel of every detective's scientific equipment.

Conan Doyle's son Adrian claimed that his father was the first to come up with the idea of using plaster of Paris to preserve footprints. This is, of course, based on Holmes's remark in SIGN:

> Here is my monograph upon the tracing of footsteps, with some remarks upon the uses of plaster of Paris as a preserver of impresses.

And perhaps there is truth in the claim that the French police studied the methods of Holmes by reading the 1906 volume *L'Oeuvre de Conan Doyle et la police scientifique au vingtieme siecle* (Green 1983, 109), or that the Egyptian police also studied Holmes's methods (Booth 1997, 208; Fido 1998, 100).

4

Sherlock Holmes

Chemist

Section 4.1 Introduction: Profound or Eccentric?

> *He is a first-class chemist.*
>
> —Young Stamford, *A Study in Scarlet*

The previous chapter discussed Sherlock Holmes as a scientifically oriented detective. He was also knowledgeable about science in general. Practically every story contains at least some mention of one of the sciences. Having explored how Holmes used science in his detective work, we now look at his interest in research and his love of things scientific. In *The "Gloria Scott"* (GLOR), one of just two of the sixty stories narrated by Holmes instead of Watson, he says, "during the first month of the long vacation. I went up to my London rooms where I spent seven weeks working out a few experiments in organic chemistry."[1] Watson tells us in *The Three Students* (3STU) that without his chemicals, Holmes was "an uncomfortable man." So there are clear indications that Holmes was devoted to science and that his first love was chemistry (see figure 4.1).

Commentators disagree about Holmes's chemistry abilities. Most praise Holmes as a chemist (see Cooper 1976; Gillard 1976; Graham 1945; Holstein 1954; Michell and Michell 1946). The most notable critic of Holmes's chemistry is Isaac Asimov. His objections are discussed in section 4.4. Dr. Watson even disagrees with himself about Holmes the chemist! Before Watson even meets Holmes, at the very outset of *A Study in Scarlet* (STUD), he is told by Young Stamford that Holmes is "a first-class chemist." Stamford then performs the historic role of introducing Holmes and Watson. It doesn't take Watson long to

[1] This quote has caused me to wonder for years about my own students' scientific efforts, or lack thereof, over Christmas break.

Figure 4.1 Sometimes Holmes would interrupt an investigation to pursue his interest in chemistry

realize that his new roommate is a unique mixture of knowledge and ignorance. When he learns in STUD that Holmes is unfamiliar with the Copernican theory and the composition of the solar system, Watson is stunned.

> HOLMES: you say we go round the sun. If we went round the moon it would not make a pennyworth of difference to me or to my work.
> WATSON: But the Solar System.
> HOLMES: What the deuce is it to me?

Holmes believes the brain has a limited capacity. Therefore useless facts like the nature of the solar system should be forgotten, lest they crowd out important things. Other fictional detectives don't necessarily agree.[2] Holmes himself appears to change his mind by the forty-seventh story, *The Valley of Fear* (VALL). In that case, he discusses Professor Moriarty's painting by Greuze with

[2] Rex Stout's fictional detective Nero Wolfe, possibly modeled after Mycroft Holmes, believes the opposite. In *Might As Well Be Dead*, he says, "The more you put in a brain, the more it will hold."

Inspector MacDonald. When MacDonald gets impatient with this tangent, Holmes remarks,

> All knowledge comes useful to the detective.

But the early Holmes believes that anyone with a large head may have a larger-than-average brain and more mental capacity. He expressed this idea about Henry Baker in *The Blue Carbuncle* (BLUE), the ninth story (see section 2.1). He also remarks in *The Five Orange Pips* (FIVE), the seventh story, about his concept of a "brain attic" that can hold only so much. Conan Doyle let Holmes accept the ideas of phrenology, the creation of the physiologist Franz Joseph Gall in the late eighteenth century (Smith 2009, 51). One of the tenets of this now-discredited theory was that the larger the brain, the more information it could hold. Thus intelligent people were those with the largest brain sizes. When the brains of well-known brilliant people, such as Albert Einstein, were measured, they were not unusually large, and phrenology began to lose credibility. Phrenology also claimed that the size and shape of the head could be used to deduce the character traits of a person. When this aspect of Gall's theory was used to infer racial superiority, phrenology "came to be reviled" (Smith 2009, 52).

Surprised by Holmes's early opinion on "brain attics," Watson decides to enumerate his abilities. The resulting list is extraordinary:

> Sherlock Holmes—his limits
> 1. Knowledge of Literature—Nil.
> 2. Knowledge of Philosophy—Nil.
> 3. Knowledge of Astronomy—Nil.
> 4. Knowledge of Politics—Feeble.
> 5. Knowledge of Botany—Variable.
> Well up in belladonna, opium, and poisons generally.
> Knows nothing of practical gardening.
> 6. Knowledge of Geology—Practical, but limited
> Tells at a glance different soils from each other.
> After walks has shown me splashes upon his trousers,
> and told me by their colour and consistence in what
> part of London he had received them.
> 7. Knowledge of Chemistry—Profound.
> 8. Knowledge of Anatomy—Accurate, but unsystematic.
> 9. Knowledge of Sensational Literature—Immense.
> He appears to know every detail of every horror
> perpetrated in the century.

10. Plays the violin well.
11. Is an expert singlestick player, boxer, and swordsman.
12. Has a good practical knowledge of British law.

Holmes seems to be interested only in things that will be of practical use to him in his profession:

> Well, I have a trade of my own. I suppose I am the only one in the world, I'm a consulting detective.

Seen in this light, Watson's list makes sense. It explains why he knows about poisons but not gardening. It explains his interest in sensational literature. What it does not explain is his willingness, even eagerness, to spend his time doing chemical experiments that have no relation to detection or crime. Watson has already confirmed Young Stamford's assessment of Holmes and chemistry. Holmes knows a lot about science in general, but his best science is clearly chemistry. Only chemistry can lure him from one of his cases. In *The Dancing Men* (DANC), Holmes wants to take a train back to London in the middle of the case because he has "a chemical analysis of some interest to finish." So we look at the chemistry in the stories first. In the final chapter, we examine Holmes and the other sciences.

As time goes on, Watson discovers that Holmes is more well-rounded than this early list suggests. No updated list is ever made. But we know that Watson changed his opinion in later adventures. One of the first things he changed was his view of Holmes the chemist. In FIVE, he recalls that his early assessment of Holmes's knowledge of chemistry was "eccentric," not "profound." Most readers feel that this false recollection indicates his new opinion of Holmes the chemist. By the time of the seventh story, FIVE, written four years after his initial list in STUD, Watson has downgraded Holmes's chemical abilities. This chapter allows us to form our own opinion of whether Holmes's chemistry knowledge was profound or eccentric.

At this point, it might be wise to mention the issue of the chronology of the sixty cases chronicled by Dr. Watson. For example, GLOR is the eighteenth story, published in March 1893. In it, we are told that Holmes worked on this case as a college student. Thus it happened earlier in his life than any other story. There is a considerable literature that attempts to assign dates during which the action in each story occurs. At least fifteen chronologies have been published (Dirda 2012, 128). Jay Finley Christ, a well-known Holmesian chronologist, claims that GLOR took place in late September of 1876. His date for FIVE is Tuesday, September 24, 1889. He chooses 1889 over 1890 because Watson remarks on the "hard rain." Christ checked the actual records of the weather office and found that hard rain occurred only in 1889. For those interested in such details,[3] Christ's work is a good place to start (Christ 1947).

Section 4.2 Coal-tar Derivatives and Dyes

I spent some months in a research into the coal-tar derivatives.
—Sherlock Holmes, *The Empty House*

In Holmes's time and well before, London's streets were illuminated by gas lamps. The gas was derived from the distillation of coal. Millions of tons of coal were processed every year to supply the gas. The coal was heated in closed vessels in the absence of oxygen. There were byproducts from this process, and they were initially considered useless. One of the byproducts was a large amount of oily tar, called coal tar. It was deemed so worthless that anyone could have it for free (Garfield 2001, 23). Gradually though, chemists were able to extract useful chemicals from the coal tar. A major step occurred in 1856 when William Henry Perkin was able to isolate a beautiful purple molecule from coal tar. The very large synthetic-dye industry arose in the years following Perkin's discovery.

During the Great Hiatus following *The Final Problem* (FINA), Holmes worked on coal-tar derivatives in Montpellier in southern France. We're never told what aspect of coal-tar derivatives was the object of his research. Moss has proposed that Holmes was attempting to isolate carcinogens from coal tar (Moss 1982, 41). Clark suggests that Holmes was active in the development of radiation technology (Clark 1964). Caplan, Inman, and the Michells disagree. They suggest that synthetic dyes were Holmes's focus (Caplan 1989; Inman 1987; Michell and Michell 1946). Stinson also supports the idea of dyes, and I agree with the majority (Stinson 2003).

At the time of Holmes's Great Hiatus (1893–1903), England was losing the industrial battle for pre-eminence in dyes. Caplan's suggestion is that Holmes was involved in a patriotic attempt to revive the English dye industry. William Henry Perkin had started the "world's first high-tech science-based industry" (Travis 2007, 43) when he accidentally created mauve, a brilliant purple dye-stuff, in 1856 in London. It is well known that he was trying to find a synthetic way of making quinine. Failing that, he pushed forward to see what was in the "reddish" and "perfectly black" powders that he had obtained instead of quinine (Garfield 2001, 36). When he extracted a beautiful purple color from

[3] Conan Doyle wasn't.

the black powder, Perkin switched gears. The eighteen-year-old student, with financial backing from his father, built a dye factory. Perkin initially called his dye Tyrian purple (Garfield 2001, 43) after the well-known natural and expensive purple dye long harvested from mollusks (particularly *Murex brandaris*) from the Mediterranean Sea. It was very expensive because it took 8,000 snails to produce one gram of Tyrian purple. Such expense gave rise to the word "porphyrogenitus," literally "born in the purple," an indicator of great wealth. Julius Caesar decreed that only the emperor and his family could wear purple garments (Garfield 2001, 39). "The mighty of the world all coveted this rare commodity" (Born 1937, 115).

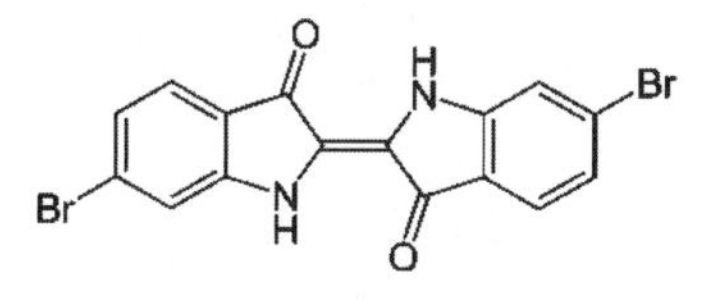

Until Perkin's discovery, clothes could only be colored using natural dyes extracted from plants (indigo) or animals (mollusks). In 1856, England was spending over £2,000,000 importing dyestuffs (Saltzman and Kessler 1991, 7). Synthetic dyes like mauve were much cheaper. Soon natural dyes would be priced out of existence. The Tyrian purple molecule is dibromoindigo, $C_{16}H_8Br_2N_2O_2$. Replacing the two bromine atoms with hydrogen atoms creates the indigo molecule, $C_{16}H_{10}N_2O_2$. Indigo has the wavelength of light shifted so that it is a blue dye. Natural indigo is extracted from a plant. Here we have one of only a handful of examples where an animal, *Murex brandaris,* and a plant, *indigofera,* produce essentially the same molecule (Hoffmann 1990, 309). Because Great Britain was importing over a million pounds of indigo each year, chemists considered the laboratory synthesis of indigo to be the "Holy Grail" (Garfield 2001, 124). No wonder chemists were hard at work on coal-tar colors! A commercial synthesis was guaranteed to be very profitable. One method for producing the dye in the laboratory went commercial in 1897. In that year, nearly 2,000,000 acres in India were used for growing indigo plants (Roberts 1989, 72). Within twenty years, the indigo crop was of no importance (Garfield 2001, 126). It could not compete with the inexpensive commercial process that produced exactly the same molecule. In the United States, indigo plants were harvested beginning in 1747, mainly in South Carolina. By the time of the American Revolution, South Carolina was exporting a million pounds of indigo per year to Europe (McKinney 2011, 4). But the American indigo crop was neglected during the war. Afterward it could not compete with India's crop, and it slowly went out of business. It finally disappeared completely after the U.S. Civil War (sciway3.net/proctor/state/sc_rice.html). Today indigo production continues to be a big business, with over 34,000,000 pounds produced worldwide in 2002.

Empress Eugenie of France, a fashion setter, began to wear clothes colored by Perkin's purple in 1857. Then, when Queen Victoria also chose to wear purple to her daughter's wedding in 1858, the popularity of the dye soon to be called mauve exploded.

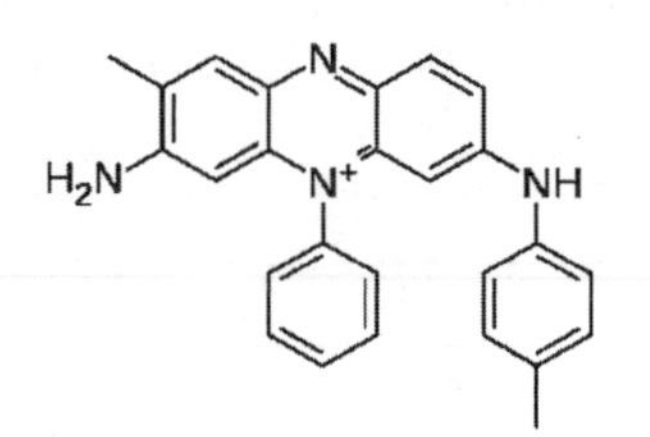

Perkin quickly became wealthy and retired from the industry at age thirty-six. Now "for the first time, people realized that the study of chemistry could make them rich" (Garfield 2001, 48). It didn't take long for people in other countries to start their own dye industries. The British scientific establishment had a great aversion to commercial aspects of their craft (*U.S. News & World Report,* April 30, 2001). But the Germans eagerly pursued the profits to be had from dyes. Soon the German dye industry surpassed that of England. The nature of patent law in the two countries favored German advances (Saltzman and Kassler 1991, 10). Fully 80 percent of dyes sold in England were being made in Germany (Garfield 2001, 146). The noted British educator and chemist Henry Enfield Roscoe lamented in 1881 (Saltzman and Kessler 1991, 9):

> To Englishmen it is a somewhat mortifying reflection that whilst the raw materials from which all these coal tar colours are made are produced in our country, the finished and valuable colours are nearly all manufactured in Germany.

The German viewpoint was different and perhaps overly enthusiastic. We can read about it in the preface to Theodore Weyl's 1885 book on coal-tar colors (Caplan 1989, 30):

> Thanks to the cooperation of theory and practice, the coal tar industry of Germany has conquered the world, and inasmuch as new and improved methods are continually being devised, will be able to maintain its pre-eminent position.

Arthur Conan Doyle would have been aware of this decline in the English dye industry. When he had Holmes work on coal-tar derivatives in France, it is likely that he had the German dominance in the coal-tar dye industry in mind. The ever-practical Holmes was doing research on dyes in an effort to stem the tide of German industry.

Section 4.3 Chemical Poisons

> *I dabble with poisons a good deal.*
>
> —Sherlock Holmes, *A Study in Scarlet*

GASES: CO AND CO_2

Today carbon monoxide, CO, is not generally thought of as a murder weapon. It is, however, still used to accomplish suicide. A closed garage with a running automobile will take only about five to ten minutes to kill anyone in the garage (Blum 2011, 134). In the early twentieth century, it was sometimes used for murder. The victim's lungs would be filled with CO. A convenient source was illuminating gas, a mixture of CO, hydrogen gas (H_2), and some hydrocarbons. Illuminating gas was initially made by the burning of coal, leaving behind the coal tar discussed in the previous section. The first house to be lighted by illuminating gas was that of William Murdoch in Cornwall, England, in 1792. Soon streets in cities were being lighted. Baltimore was the first American city to have lighted streets. They did so with illuminating gas beginning in 1821. With this deadly mixture of gases now available in houses, some deaths were bound to occur. Accidents, suicides, and murders were the result.

There are four deaths by asphyxiation in the Canon. One is the hanging of Blessington in *The Resident Patient* (RESI). The other three involve oxygen deprivation. The most clear-cut of these occurs in *The Greek Interpreter* (GREE). Paul Kratides is being held by Harold Latimer and Wilson Kemp, who are attempting to get him to sign over valuable property. Because Kratides speaks no English, the crooks bring Mr. Melas to ask questions in Greek of their captive. Melas is a well-known linguist who is frequently hired to interpret, particularly in his native tongue, Greek. Shortly into the interview with Kratides, Melas devises a way to find out what is really going on. He begins to add his own questions to the words of the criminals. He asks their question immediately followed by one of his own. Kratides responds to both. As everything is spoken in Greek, Latimer and Kemp don't have a clue about what is happening.

MELAS: You can do no good by this obstinacy. *Who are you?*
KRATIDES: I care not. *I am a stranger in London.*
MELAS: The property can never be yours. *What ails you?*
KRATIDES: It shall not go to villains. *They are starving me.*

At the end of the interview, they set the interpreter free. Having realized that a crime is in progress, Melas consults his acquaintance from the boarding house where he lives, Mycroft Holmes. When Mycroft takes the case to Sherlock, a surprised Watson learns that Sherlock has a brother. When Holmes gets a lead about the whereabouts of Kratides, he goes to collect Melas to help interpret again. But Melas has been kidnapped by the thugs. Then Melas is left with Kratides as Latimer and Kemp run off with Kratides's sister Sophy, who has fallen under the influence of Latimer.

Holmes, Watson, and Inspector Gregson find Melas and Kratides tied up in a room with charcoal burning in a small brass tripod. The incomplete combustion has resulted in production of carbon monoxide. Soon the amount of CO is sufficient to kill the weakened Kratides. Melas just barely survives due to the timely arrival of Watson, who administers first aid (see figure 4.2). The killers escape but soon meet their ends in Hungary.

Figure 4.2 Melas survives but Kratides succumbs to carbon-monoxide poisoning.

Carbon-monoxide poisoning occurs because the Fe in hemoglobin bonds 200 times more strongly to CO than it does to O_2. Thus when both gases are present, it is mainly CO that attaches to the Fe in hemoglobin (Blum 2010, 137). In this way, the blood circulating to the brain carries too little oxyhemoglobin and too much carboxyhemoglobin. The result is suffocation due to lack of O_2. The blood turns cherry red due to the carboxyhemoglobin (Curjel 1978, 155). The case is not one of Holmes's great successes. It may not have been one of Conan Doyle's either. Kratides and Melas are described as "blue-lipped," a coloration associated with cyanide poisoning, not carboxyhemoglobin.

Though GREE has an entertaining plot, the story is overwhelmed by the appearance of Mycroft Holmes. The surprise of his existence, not revealed until the twenty-fourth story, and the vivid characterization that Conan Doyle presents, tend to divert the reader's attention from the actual story. We learn much of Mycroft's fascinating background and character from this story.

Unlike the situation in GREE, the other suffocations take place in confined spaces, that is, places where the replenishment of oxygen is hindered or stopped. In such spaces, deprivation of oxygen necessarily results. In *The Retired Colourman* (RETI), murder is committed by suffocation using an unnamed gas in a "hermetically sealed room." Perhaps it too was carbon monoxide (Campbell 1983, 19). Illuminating gas could have been the source of the CO. Josiah Amberly kills his young wife and her lover. He then consults Holmes to solve the "disappearance" of his wife. This is a big mistake. As Holmes says, "He felt so clever and so sure of himself that he imagined no one could touch him." Fittingly the smell of another chemical shows Sherlock the truth. Amberly is the "colourman" mentioned in the story's title. Even so, why would a distraught husband be painting the inside of his house at this time? Holmes deduces that Amberly wasn't distraught; he was using the strong odor of the paint to cover up the smell of the gas.

The third suffocation occurs in *The Musgrave Ritual* (MUSG), a case brought to Holmes by one of his few college friends, Reginald Musgrave. It involves recovery of the long-lost ancient crown of the king of England. The location of the crown is described by the ritual. It is first solved by Richard Brunton, the Musgraves' butler for twenty years. When Brunton is discovered inappropriately looking at family materials at 2 AM one Friday morning, the enraged Reginald Musgrave fires him and gives him a week's notice. Brunton disappears the next Sunday morning. Holmes is called in to investigate the following Thursday. He also unravels the cryptic directions to the location of the crown that are given in the ritual:

> "Whose was it?"
> "His who is gone."
> "Who shall have it?"

"He who will come."
"Where was the sun?"
"Over the oak."
"Where was the shadow?"
"Under the elm."
"How was it stepped?"
"North by ten and by ten, east by five and by five, south by two and by two, west by one and by one, and so under."
"What shall we give for it?"
"All that is ours."
"Why should we give it?"
"For the sake of the trust."

Following the directions given in the Musgrave ritual (see section 5.1) leads Holmes to a small cellar room, 4 feet square and 7 feet high. The large stone slab covering the top of the room must be larger than 4 feet by 4 feet. Holmes needs help from a burly policeman to remove it. But the room contains no crown, just Brunton's dead body.

Brunton's accomplice in this caper is Rachel Howells, the Musgraves' maid, to whom he was formerly engaged. Inexplicably Brunton turned to the woman he scorned for help in his plan to steal the goods. After they leveraged the slab up and supported it with a 3-foot-long billet of wood, Brunton descended into the chamber and passed the treasure up to Howells. Once he handed up the crown, she kicked the support away, the large stone fell back in place, and Brunton was left to suffocate (see figure 4.3). Presumably this occurred in the early-morning hours of Sunday, since Brunton's absence began Sunday morning. There are no indications of foul play. Brunton died from CO_2 poisoning.

The scorned Howells, "of Welsh blood, fiery and passionate," killed Brunton, threw the treasure in the nearby lake, and disappeared three days later. When Conan Doyle wanted to use the stereotype of a hotheaded woman, he usually turned to those of "tropical blood," such as the Brazilians Maria Gibson in *The Problem of Thor Bridge* (THOR) and Isadora Klein in *The Three Gables* (3GAB), the Peruvian Mrs. Ferguson in *The Sussex Vampire* (SUSS), and the Costa Rican Beryl Stapleton in *The Hound of the Baskervilles* (HOUN) (Jann 1995, 109). After all, "Englishwomen, particularly those of the higher classes, exercise more control" (Jann 1995, 109). Apparently not Welshwomen, though.

Upon finding the body, Holmes immediately declares that he had been dead for "some days." He does not indicate how he arrived at this estimate. We shall do an approximate calculation to see if Holmes's statement is even reasonable. We can compute the amount of oxygen in the room, how much Brunton would consume per hour by breathing, and then how long it would take for the percent

Figure 4.3 Sherlock Holmes and Reginald Musgrave find the dead body of Brunton the butler.

O_2 to drop to a dangerous level. The website of the U.S. Occupational Safety and Health Administration (OSHA) (http://www.osha.gov/pls/oshaweb/owadisp.show_document?p_id=25743&p_table=INTERPRETATIONS) states that an oxygen level of 19.5 percent is safe for humans and that at levels below 16 percent, deleterious health effects begin. OSHA and several other sources claim that, at 6 percent oxygen, death quickly follows. The approximate calculation below assumes that once the slab of rock is in place, no additional oxygen enters the chamber. Thus it computes the shortest time that Brunton could have survived in the small cellar before the O_2 level reached 16 percent and 6 percent.

Step 1. The volume of the room.

$$4 \text{ feet} \times 4 \text{ feet} \times 7 \text{ feet} = 112 \text{ cubic feet or } 112 \text{ ft}^3$$

Some air will be displaced in the room by Brunton's body and anything else in there. So we will estimate this to reduce the volume of air to 110 ft^3.

The computation is simpler[4] if the volume is in liters. So we will convert ft^3 to L.

$$(110\ ft^3)(12\ inches/1ft)^3(2.54\ cm/1\ inch)^3(1\ L/1000cm) \sim 3115\ L.$$

Step 2. The amount of oxygen in the room when Rachel Howells seals it.
To compute how much oxygen is in the room, we use the Ideal Gas Law, a very good approximation for normal conditions.

PV = nRT,
Where P is the pressure in atmospheres
V is the volume of the room in liters
n is the number of moles of the gas
R is the gas constant, 0.0821 l-atm/mole degree
T is the temperature in degrees Kelvin

The ambient pressure will be the normal everyday value of 1 atmosphere. But oxygen is 21 percent of air, so the pressure of O_2 at the start will be 0.21 atmospheres.
We'll use a typical summertime value for the temperature of 293°K, or 68°F. Then

$$n_{O2} = (0.21\ atm.)(3115\ L)/\ (0.0821\ L\text{-}atm/mole\ ^\circ K)(293^\circ K)$$
$$n_{O2} = 27.2\ moles$$

Brunton has 27.2 moles of O_2 when Howells seals him in the room. Now his breathing starts converting the oxygen to carbon dioxide, CO_2.

Step 3. How many moles of O_2 are left in the room when it becomes unhealthy?

When O_2 was down to 16% of the 3115 liters in the room, we find:

$$n_{O2} = (0.16\ atm.)(3115\ L)/\ (0.0821\ L\text{-}atm/mole\ ^\circ K)(293^\circ K)$$
$$n_{O2} = 20.7\ moles$$

If we subtract this amount from the number of moles of O_2 present at the beginning of his confinement, we find that when 27.2 – 20.7 = 6.5 moles of O_2 consumed, Brunton's health is affected.

[4] For the sake of simplicity, we shall ignore the number of significant figures in this calculation.

When O_2 was down to 6%,

$$n_{O2} = (0.06 \text{ atm.})(3115 \text{ L})/(0.0821 \text{ L-atm/mole °K})(293\text{°K})$$
$$n_{O2} = 7.8 \text{ moles}$$

Thus when 27.2 – 7.8 = 19.4 moles of O_2 had been consumed, Brunton is surely dead.

Step 4. How much O_2 is consumed by each breath?
The average human breath is about 0.5 liters. As stated above, the air inhaled is 21 percent oxygen, at least at the start. The basis for mouth-to-mouth resuscitation is that exhaled air also contains oxygen. So we must take into account that exhaled air is 15 percent oxygen.[5]
The number of moles of O_2 inhaled is

$$n_{O2} = (0.21 \text{ atm.})(0.5 \text{ L})/ (0.0821 \text{ L-atm/mole °K})(293\text{°K}) = 0.00437 \text{ moles } O_2 / \text{breath}$$

The number of moles of O_2 exhaled out is

$$n_{O2} = (0.15 \text{ atm.})(0.5 \text{ L})/ (0.0821 \text{ L-atm/mole °K})(293\text{°K}) = 0.00312 \text{ moles } O_2 / \text{breath}$$

Thus each breath consumes approximately 0.00437 – 0.00312 = 0.00125 moles of O_2

Step 5. How many hours before the danger levels of 16 percent O_2 and 6 percent O_2 are reached?
To consume 6.5 moles of O_2 would take

(0.00125 moles/breath)(12 breaths/minute) = 0.015 moles/minute
6.5 moles/0.015 moles/minute = 430 minutes or 7 hours 15 minutes

Brunton would be in danger by 9 or 10 AM Sunday.
To consume 19.4 moles of O_2 would take

(0.00125 moles/breath)(12 breaths/ minute) = 0.015 moles/minute
19.4 moles/0.015 moles/minute = 1,300 minutes or 21 hours 40 minutes

Brunton would be dead by midnight Sunday.

[5] An average based on information from several websites.

This approximate calculation can be improved by taking into account the fact that as the oxygen in the cellar chamber diminishes, Brunton may very well consume less of it per breath. Doing this pushes the time for reaching the dangerous 16 percent level back at most to 1 PM Sunday. The 6 percent level would surely be reached by noon Monday at the latest. Holmes found the dead body on Thursday. So his assertion that Brunton had been dead for "some days" is accurate.

PRUSSIC ACID, HCN

Prussic acid is the historical name for hydrogen cyanide, HCN. It is a deadly liquid that is most poisonous when its vapor is inhaled. It acts on a victim by interrupting cellular respiration. Thus the cyanide ion CN^-, like CO, also deprives the victim of oxygen, O_2. But the much greater toxicity of CN^- is due to its action by a mechanism different than that of CO (Greenwood and Earnshaw 1984, 1279). Cyanide poisoning is characterized by a blue tint to the skin and the well-known odor of almonds. At the end of *The Veiled Lodger* (VEIL), Eugenia Ronder gives a bottle of Prussic acid to Sherlock Holmes. He is pleased that he has dissuaded such a "brave woman" from suicide.

Eugenia Ronder and her lover, Leonardo the strongman, worked in her husband's wild beast show. Their plan to kill Mr. Ronder went awry when the lion, Sahara King, escaped from his cage. The lion's claws left Eugenia with a face that Watson described as "a grisly ruin." In the seven years since the attack, this formerly beautiful woman has worn a veil to cover her maimed face. Leonardo deserted her immediately, as she was no longer beautiful. Now that Leonardo has died, she confers with Holmes to clarify the events of that night. He senses that she might be considering suicide and encourages her, saying, "Your life is not your own." Holmes is gratified when she sends him the bottle of Prussic acid.

CHLOROFORM, $CHCL_3$

Chloroform is viewed today as an early anesthetic. It was not always so. *The Poisoner's Handbook* (Blum 2010) recounts the early history of chloroform as a poison and even a murder weapon. In 1911 on Long Island, a father used it to kill his son and two daughters and then walked into Atlantic Ocean to his own death. Another early gruesome use of chloroform to kill occurred in 1915 in Yonkers, New York. Frederic Mors used the readily available $CHCl_3$ to deliberately kill elderly pensioners at the German Odd Fellows home. Mors willingly carried out the wishes of the superintendent of the facility to perform these "mercy killings." He would administer whatever dose of chloroform was necessary (Blum 2010, 7).

But the designed purpose of chloroform was as an anesthetic. James Simpson in Edinburgh deliberately inhaled chloroform to see if it had anesthetic properties. He and two assistants spent the evening of November 4, 1847, inhaling such molecules as acetone and benzene to check their anesthetic properties. None worked until they tried chloroform. It was so effective, and they recovered so well that Simpson thought, "This will change the world" (Blum 2010, 10). However, by the start of the twentieth century, the British Medical Association called chloroform "the most dangerous anesthetic known." Still, its use as an anesthetic persisted for years.

Conan Doyle employs $CHCl_3$ in three Holmes stories. Its use never results in a death. In 3GAB, Barney Stockdale is hired by Isadora Klein and uses chloroform to subdue Mrs. Maberly in order to steal a manuscript written by her son Douglas. Isadora Klein is intent on obtaining the manuscript because its publication will reveal her past and surely result in cancellation of her upcoming marriage to the young Duke of Lomond. By the time Holmes arrives at her home, she has burned the manuscript. This prevents Holmes from returning the manuscript to Mrs. Maberly. Instead, as recompense, he persuades Isadora Klein to bankroll a first-class trip around the world for Mrs. Maberly.

In *His Last Bow* (LAST), Holmes plays the role of a double agent. He appears to be working to obtain British naval secrets for the German spy Von Bork. He has convinced Von Bork that he is an American named Altamont. He arrives, driven by his chauffeur, to deliver the material. Watson is the chauffeur, and the two of them subdue the German. Von Bork is chloroformed and captured (see figure 4.4). LAST was written in 1917 near the beginning of World War I.

The most dramatic use of chloroform is in *The Disappearance of Lady Frances Carfax* (LADY). This is an instance where the loyal Watson travels to investigate the facts. Five weeks have passed since Lady Frances last wrote home from Lausanne, Switzerland. Watson tracks her to the Englisher Hof in Baden. Henry Peters of Adelaide steals Lady Frances's jewels, carries her off to his London home, and then tries to bury her alive in a doubly occupied coffin. She is chloroformed and kept that way. Holmes picks up the trail when her jewels are pawned. He arrives in time to figure out the double- coffin trick, but not in time to catch Peters.

Other poisons are mentioned throughout the sixty Holmes stories. Despite the fame of arsenic and its nickname, "inheritance powder," Conan Doyle never used it in any Holmes case. Aqua Tofana, an arsenic concoction of seventeenth-century Italy, receives a brief mention in the first story, STUD. In seventeenth-century Naples (Klinger 2006, vol. 3, 93) a woman named Tofana or Teofania di Adamo (Wagner 2006, 47) used it to commit over six hundred murders. When she came under suspicion, she took

Figure 4.4 The spy Von Bork is subdued by chloroform in His Last Bow.

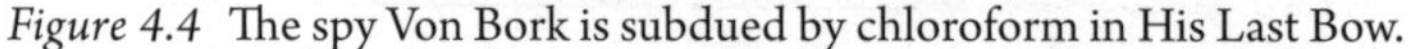

refuge in a convent. Subsequently expelled by the sisters, she confessed under intense questioning to the murders. It is said that she soon died of strangling (Wagner 2006, 47). Most other poisons in the Holmes stories are what Conan Doyle calls "vegetable alkaloids." We discuss them in the biology section in the final chapter.

Section 4.4 Asimov's View: Holmes the Blundering Chemist

A remarkable worm, unknown to science

—Isaac Asimov's Investiture in the Baker Street Irregulars

Isaac Asimov was a chemistry professor, a prolific writer, a hugely popular speaker, and a Sherlockian scholar. In 1980, he attacked the chemical knowledge of Sherlock Holmes, calling him the blundering chemist (Asimov 1980). In a 1983 introduction to "Sherlock Holmes on Medicine and Science" (Simpson

1983), Asimov tried to shift the blame for what he deemed to be Holmes's chemical deficiencies first to Dr. Watson and then to Arthur Conan Doyle. I intend to show that perhaps there is no blame to be shifted. It may just be that Asimov missed a point or two in his analysis. Perhaps Holmes the chemist made no blunders. I discuss three major points of Asimov's criticism: acetones, gemstones, and the Sherlock Holmes blood test.

ACETONES

In *The Copper Beeches* (COPP), Violet Hunter is offered a position as a governess at a country estate named The Copper Beeches. Before accepting, she consults Sherlock Holmes because several factors have aroused her suspicions. First, her salary is to be two and a half times larger than that in her previous position. Also, her employer, Jephro Rucastle, will provide an "electric blue" dress that she is to wear when asked. And, worst of all, she must cut her "luxuriant" chestnut hair, of which she is very fond, short. Holmes confesses that it "is not the situation which I should like to see a sister of mine apply for." When Miss Hunter reminds him of the salary, he says, "[T]he pay is good,—too good."

Violet Hunter resolves to accept the position of governess to six-year-old Edward Rucastle. Holmes tells Watson, "I am much mistaken if we do not hear from her before many days are past." As he waits to hear from the governess, Holmes settles down "to one of those all night chemical researches." But when Violet Hunter's telegram arrives late one night Holmes says,

> Perhaps I had better postpone my analysis of the acetones.

Asimov points out that there is just one molecule named acetone. It is not the name of a class of molecules. Every chemist knows this. So the fact that Holmes doesn't suggests that he is incompetent in chemistry.

Acetone belongs to the class of molecules called ketones. All ketones have this structure[6]:

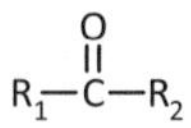

$$R_1-\overset{\displaystyle O}{\overset{\|}{C}}-R_2$$

Ketones differ from one another by having different molecular fragments R_1 and R_2. These fragments are generally hydrocarbon pieces containing different

[6] In general, chemical structures, including the ones we discuss, are not planar.

numbers of carbon atoms, such as CH_3, C_2H_5, C_3H_7, and larger. Acetone is the simplest (i.e., the smallest) ketone because both R_1 and R_2 have only one carbon atom, and both are methyl groups, CH_3. Thus the chemical formula of acetone is $CH_3(CO)CH_3$, and its chemical structure is:

```
    H   O   H
    |   ||  |
H — C — C — C — H
    |       |
    H       H
```

Asimov is certainly correct that, in today's usage, the word "acetone" is not used as the name of a class of molecules. But it was different in Holmes's time. In Adolph Strecker's *Short Textbook of Organic Chemistry* (cited in Redmond 1964), one finds that the usage was different then. We read:

> By replacement of two hydrogen atoms of a paraffin on one and the same carbon atom, there result derivatives... whose oxygen compounds are termed ketones or acetones.

A paraffin is a molecule containing only hydrogen atoms and carbon atoms (i.e., a hydrocarbon). It is the type of hydrocarbon in which all the chemical bonds in the molecule are single bonds. For that reason, it is said to be "saturated," and it has the maximum number of hydrogen atoms for the number of carbon atoms. Following the prescription in the above quote, if the two hydrogen atoms on the central carbon atom of the "paraffin" propane, C_3H_8, are replaced by an oxygen atom (which, unlike hydrogen, can form a double bond to carbon), the result is acetone.

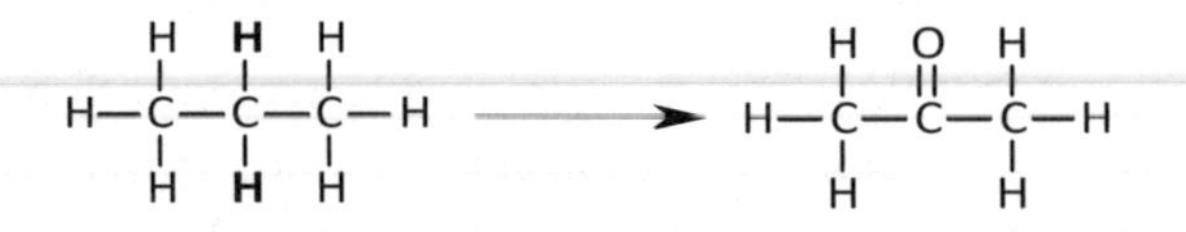

That an entire "class" of molecules can be formed this way is clear if we consider the next larger paraffin or hydrocarbon, butane C_4H_{10}. The ketone that would result upon replacement of two hydrogen atoms by one oxygen atom has one methyl group, CH_3, and one ethyl group, C_2H_5. It is called methyl ethyl ketone (MEK), $CH_3(CO)C_2H_5$.

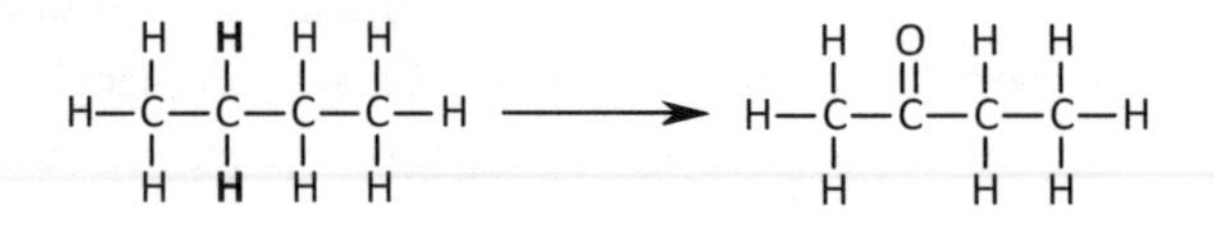

The series continues for hydrocarbons with more carbon atoms.

The entire class of molecules used to be referred to as "ketones or acetones" (Wislicenus 1885, 275). So in Sherlock Holmes's world of chemistry, it was perfectly acceptable to say "acetones" to indicate the whole set of molecules that we now term ketones. All chemists of Holmes's era would have understood him perfectly. Asimov apparently didn't research old-fashioned chemical nomenclature.

As to Violet Hunter's problem, Holmes answers her summons and arrives in Hampshire with "seven separate explanations, each of which would cover the facts as we know them." Violet Hunter then describes her experience as a governess at The Copper Beeches. In addition to the blue dress and the shorn hair, she must sit with her back to a picture window and listen to the normally taciturn Jephro Rucastle tell "uproariously funny jokes." During one such session, she manages to discern a young man in the street watching this performance. Armed with this additional information, Holmes chooses one of his seven theories and says,

> [T]here is only one feasible explanation. You have been brought there to personate someone.

That someone is Alice Rucastle who has not gone to Philadelphia, as Violet was told, but is imprisoned in the attic. Jephro Rucastle wants to discourage Mr. Fowler, Alice's suitor in the street, from coming round by showing him that Alice (actually her look-alike Violet Hunter) is quite happy without him. In this way, the Rucastles hope to keep Alice unmarried so that they will retain control of her money. Fortunately both Violet Hunter and the imprisoned Alice Rucastle escape the plans of Jephro Rucastle and go on to better things.

It has been noted that COPP has a host of similarities to Charlotte Brontë's *Jane Eyre* (Duyfhuizen 1993). These stories, both with governesses as the main character and a woman imprisoned in an attic, are about the independence and empowerment of women. This theme "was still largely atypical for 1891" (Duyfhuizen 1993, 143). Surely the well-read Conan Doyle was familiar with Brontë's 1847 novel, and it may well have influenced the plot of COPP.

GEMSTONES

Isaac Asimov considered Sherlock Holmes's knowledge of gemstones to be deficient. He based this conclusion on several Holmes comments in BLUE. When Watson asks if the gemstone that has come into their possession is the Countess of Morcar's missing blue carbuncle, Holmes responds,

> Precisely so. I ought to know its size and shape.

Asimov rightly points out that any competent chemist should know that carbuncles are never blue. The red almandine garnet has the chemical formula $Fe_3Al_2(SiO_4)_3$ (Rutland 1974, 185). It is the stone that is also known as a carbuncle (Sinkankas 1962, 99). Then Holmes makes it worse by referring to "the precious stone" as "crystallized charcoal." Now Asimov claims that Holmes is confusing a carbuncle with a diamond.

There have been several attempts to explain Holmes's statements. For example, Redmond mentions that Watson, the chronicler, may have deliberately misnamed the gem (Redmond 1964, 151). This seems like an unsatisfactory explanation. Bigelow refutes Beckemeyer's claim that the gem was a blue sapphire. He says it is a blue diamond (Bigelow 1961, 212), and that the countess called it a carbuncle out of ignorance or whim. Kasson (1961) agrees with Bigelow and identifies exactly which blue diamond, the famous Hope diamond. So does Hunt (2011), except he says the carbuncle is actually the Brunswick blue diamond. Redmond considers the Hope as a plausible candidate. Waterhouse (2004) opts for "a large flawless cobalt blue spinel." Blank (1947, 237) supports Asimov and states that Holmes had a "deplorable lack" of knowledge when he said the countess's carbuncle was crystallized charcoal. Confusion reigns. Is the "blue carbuncle" a diamond, a sapphire, a spinel, a carbuncle, or some other gemstone? All of these explanations require that someone, Watson, Holmes, or the Countess of Morcar, made a mistake.

Another explanation is that no one erred. The gem was most likely a doublet. Doublets, made since Roman times, were encountered extensively in Victorian jewelry (Rutland 1974, 56). The purpose of creating a doublet was to enhance the size and appearance of a stone, or to imitate a more valuable gem. Doublets consisted of a gemstone, most often a garnet, fused to the top of a stone that was generally of lesser value, frequently glass. Garnets were the gem of choice for the top of doublets because they retained their luster and durability and did not crack upon fusion (see figure 4.5). By adding a thin portion of red garnet, "any colored gem could be simulated" (Matlins and Bonanno 1993, 138). By adjusting the thickness of the garnet on top, the red color would not be seen. Thus we begin to see how the confusion arose. If a carbuncle was used in a doublet to produce a blue color, it would be easy enough to refer to it as a blue carbuncle.

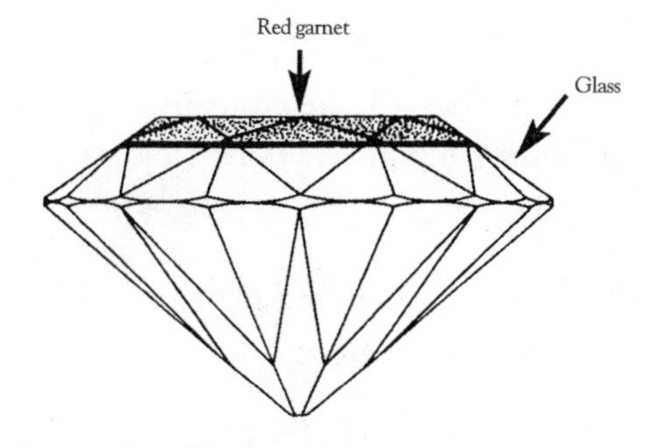

Figure 4.5 Garnet-topped doublet

Souce: Jewelry & Gems, The Buying Guide 7 Edition: How to Buy Diamonds, Pearls, Colored Gemstones, Gold & Jewelry with Confidence and Knowledge 2009 by Antoinette L. Matlins PG & Antonio C. Bonanno, FGA, P, ASA. Permission granted by GemStone Press, www.gemstonepress.com

The most reliable way of detecting the presence of a doublet is to immerse it in rubbing alcohol. However, this method does not work with garnet-topped doublets (Matlins and Bonanno 1989, 176). Knowing that the refractive index varies from one gemstone to the next, a person might very well decide to measure this property. This is a physical test that shines light on the substance being tested. Light travels more slowly when going through materials, particularly solids and liquids. This gives rise to the phenomenon of a fishing line appearing to bend as it enters the water, even though it is actually straight. The effect can be measured quantitatively, and each substance has its own value. The refractive index, RI, is the ratio of the speed of light in a vacuum over its speed in the substance. The RI value of carbuncles is 1.76 to 1.83 (Matlins and Bonanno 1989, 108). In testing a doublet, if the test light was shone on the thin carbuncle layer, the resulting RI value would be that of the carbuncle. Thus a stone that appeared blue and had a refractive index in the range 1.76–1.83 might well be called a blue carbuncle, even though the bulk of the stone was not a carbuncle but some other substance.

But what are we to make of Holmes's remark about crystallized charcoal? Someone as wealthy as the Countess of Morcar would have no need to attach her garnet to glass. Remember, we have Holmes's word that "It is absolutely unique." So we must consider the possibility that he was right again, and that the bottom of the doublet was actually a diamond. Diamond doublets are not often encountered, but they do exist (Matlins and Bonanno 1989, 171). They usually consist of two small diamonds glued together to make a larger stone. It appears that Holmes is telling us that the Countess of Morcar's famous gemstone is indeed unique, with a diamond bottom and a carbuncle top. We cannot call Sherlock Holmes a blundering chemist when so logical an explanation of his comments is possible.

THE SHERLOCK HOLMES BLOOD TEST

Recall that in STUD, Young Stamford takes Watson to the laboratory at St. Bart's Hospital in order to introduce him to Sherlock Holmes. As they enter the lab, before any introduction, they hear Holmes calling out,

> I've found it! I've found it! I have found a re-agent which is precipitated by haemoglobin, and by nothing else.

After being introduced to Dr. Watson and perceiving that he'd been in Afghanistan, Holmes asks him what he thinks of the Sherlock Holmes blood test. Watson responds,

> It is interesting chemically no doubt, but practically...

The excited Holmes interrupts before Watson can finish his criticism:

> Why, man, it is the most practical medico-legal discovery for years. Don't you see that it gives us an infallible test for blood stains?

Asimov does not question the existence or the effectiveness of the Sherlock Holmes blood test. But he doubts that it is as sensitive as Holmes claims. Using Holmes's description, Asimov computes that the relative volumes of blood and water in his test are 1 to 50,000 (Asimov 1980, 12). Yet Holmes states, "The proportion of blood cannot be more than one in a million." A good chemist, says Asimov, would get closer to the truth than that and "could not possibly make this mistake."

A potential source of error in Asimov's calculation is that, in Europe, quantities were, and still are, calculated in terms of weight rather than volume. This is particularly true for recipes, but it used to be somewhat true for scientists (wikipedia.org/wiki/Apothecaries).

Is the proportion of blood in water anywhere near one in a million as Holmes claims? Using the same dilution factor as Holmes and Asimov, one can compute a weight/weight ratio close to 1 in a million. The calculation uses the fact that 1 gram of water is also 1 milliliter. It also assumes that there are 5 grams of hemoglobin (Hb) in 100 milliliters of blood. This is not quite right. The amount of hemoglobin in 100 milliliters of blood is closer to 15. But for a calculation done in his head 125 years ago, Holmes did well.

Asimov [1 in 50,000]
0.02 ml blood/1000 ml H_2O = 1 ml blood/50,000 ml H_2O
O'Brien [1 in 1,000,000]
[5 g Hb/100 ml blood][.02 ml blood/1000 g H_2O] = 1g Hb/10^6 g H_2O

It is clear that Asimov may have been overly harsh in judging Sherlock Holmes the chemist.

Holmesian scholars have written numerous times on the Sherlock Holmes blood test. A good review of the history of blood testing in the nineteenth century has been given by McGowan (1987). There we learn that a variety of chemicals had been used prior to Holmes's time to detect blood. The early nineteenth-century tests of Barruel in 1829 and Bryk in 1858 both used concentrated sulfuric acid as the test reagent. Teichmann's test of 1853 used glacial acetic acid and sodium chloride. In 1861, the Van Deen test used guaiacum followed by turpentine or hydrogen peroxide. This is probably the test Holmes refers to when he tells Watson at their first meeting,

> The old guaiacum test was very clumsy and uncertain.

Two tests developed in the 1870s were the Zahn test, which used hydrogen peroxide, and the Sonnenschein test, which used sodium tungstate and acetic acid. Even in 1911, Britain was using a test with turpentine and benzedrine or guaiacum to look for a blue coloration (Fido 1998, 100). So it seems that his test was not enough of an improvement to be put into general use.

Huber (1987) has produced the best candidate for Holmes's test and shows that it was still in use a full century after Holmes made his discovery in STUD. Her candidate for the Sherlock Holmes blood test is the addition of sodium hydroxide followed by saturated ammonium sulfate.[7] She notes that this test does not distinguish human blood from animal.

Section 4.5 Other Chemicals

> Watson: *Well, have you solved it?*
> Holmes: *Yes. It was the bisulphate of baryta.*
> Watson: *No, no, the mystery.*
>
> *A Case of Identity*

BARIUM BISULPHATE

There were a number of other chemicals that played a lesser role in the Holmes tales. In *A Case of Identity* (IDEN), Watson, now married to Mary

[7] "Saturated" means that the maximum amount possible of ammonium sulfate has been dissolved in water.

Morstan and no longer living with Holmes, returns to Baker Street and finds that Holmes has spent the day working on a chemical analysis. He is so intent on his chemical results that when Watson asks if he'd solved the mystery, Holmes mistakes the inquiry for a question about his chemical work. Once again, as in the case of the "acetones," Holmes uses old nomenclature. Baryta is a no-longer-used term for barium oxide, BaO. Thus the "bisulphate of baryta" is barium bisulphate, $Ba(HSO_4)_2$. Asimov has mild criticism for Holmes's use of the term, stating that he should simply have said barium bisulphate. He also claims that it is not particularly difficult to analyze. That's certainly correct. But the problem with $Ba(HSO_4)_2$ is not in the analysis. The problem is obtaining it in the first place. Barium bisulphate is rare enough to have been called a "curiosity" (Klinger 2006, vol. 1, 92). In fact, the very existence of the compound has been called into question (Tracy 1977, 27).

The famous Swedish chemist Berzelius first claimed to have isolated it in 1843 (Gillard 1976, 11). He mixed sulfuric acid, H_2SO_4, with barium sulphate, Ba_2SO_4, and by cooling the mixture detected $Ba(HSO_4)_2$. But the molecule continued to be elusive. A 1921 study of the freezing points of mixtures of H_2SO_4 and Ba_2SO_4 showed no signs of the bisulphate (Kendall and Davidson 1921). Finally in 1931, solubility and conductivity studies firmly established the existence of $Ba(HSO_4)_2$ (Trenner and Taylor 1931). Additional studies over the years have verified that barium bisulphate definitely does exist (Hammett and Lowenheim 1934). Once again, we find Holmes at the forefront of nineteenth-century chemistry, working with a substance that real chemists were finding difficult.

HYDROCARBONS

In *The Sign of the Four* (SIGN), Holmes spends time in the middle of the case doing some chemistry: "When I had succeeded in dissolving the hydrocarbon which I was at work at." Asimov dismisses this as a trivial experiment (Asimov 1980, 16). Other Sherlockians agree, as evidenced by their efforts to think of some way this work could be chemically significant. One suggestion is that Holmes had a mixture of hydrocarbons (Redmond 1964, 145). That Holmes does not name any particular hydrocarbon lends credence to this suggestion. Still, a mixture would not be significantly more difficult to dissolve than a single hydrocarbon. Redmond tries to increase the importance of Holmes's hydrocarbon work by suggesting that it was a preliminary step in a more important task, most likely forensic in nature.

Two other scholars claim that Holmes was working on a specific hydrocarbon. Cooper (1976, 71) states that he was not particularly trying to dissolve the hydrocarbon, but was instead characterizing it by measuring its physical characteristics. Walters (1978, 223) actually identifies which hydrocarbon was the subject Holmes's efforts, a hydrocarbon-like molecule, carbazole. The claim is made that Holmes succeeded in dissolving it in sulfuric acid, H_2SO_4, an unusual solvent for hydrocarbons.

The best conclusion is that Holmes was not doing important chemistry in dissolving a hydrocarbon.

ACIDS

There are a number of acids mentioned in the sixty stories. Several times we hear of Holmes leaving acid stains in the Baker Street rooms. But he did not often use acids in his forensic work. Watson tells us about one important case where an acid test is used to prove a man's guilt. It happens in *The Naval Treaty* (NAVA):

> You come at a crisis, Watson. If this paper remains blue, all is well. If it turns red, it means a man's life.

When the litmus paper turns red, the unsurprised Holmes fires off several telegrams to the authorities. Alas, this is one of the untold tales. It is not a part of Holmes's investigation of the theft of the naval treaty. We know nothing else at all about the case, or the chemistry used to prove guilt.

Another acid we encounter is carbolic acid, C_6H_5OH, which is also called phenol. It was discovered in 1834 when it was extracted from coal tar. The famous Sir Joseph Lister made use of its antiseptic properties (Klinger 2005, vol. 1, 267). These same properties are why Conan Doyle mentions it in two Holmes cases, *The Cardboard Box* (CARD) and *The Engineer's Thumb* (ENGR). In ENGR, it is Watson who uses carbolic acid to dress Victor Hatherly's mutilated thumb. The carbolic acid plays no significant role in the case, although it was discussed when Watson's skill as a physician was examined in section 2.2.

CARD has been termed "easily the darkest tale in the entire Canon" (Klinger 2005, vol. 1, 422). Jim Browner is married to Mary Cushing. But her sister Sarah has designs on Jim. When he rejects her, Sarah proceeds to turn Mary against her husband. When he catches his wife with another man, the hotheaded Jim follows them, kills them both, and cuts off an ear from each. He then mails the two severed ears to Sarah to show her what she has caused. Early on, Inspector Lestrade mentions the possibility of a medical-student prank. But Holmes realizes that the fact that carbolic acid was not used to preserve the two ears means that they were probably not sent by a medical person.

Sulfuric acid is also employed in two stories, BLUE and *The Illustrious Client* (ILLU). Both instances involve vitriol (an old name for sulfuric acid, H_2SO_4) throwing. In BLUE, we learn nothing except that a vitriol-throwing incident is part of the unhappy history of the Countess of Morcar's blue carbuncle.

In ILLU, the acid does play an important part in the plot. Baron Adelbert Gruner, called by Holmes "the Austrian murderer," has won the heart of the beautiful Violet de Merville. An unnamed "illustrious client" engages Holmes to convince Violet of Gruner's true character. She is immune to persuasion despite all the evidence that Gruner has cast a number of women aside after cruelly using them. One of his past mistresses, Kitty Winter, is full of hate for the handsome Gruner. Holmes takes Kitty to talk to Violet. That strategy also fails. But Kitty Winter has her own plan for revenge. She takes some sulfuric acid and throws it in evil Baron Gruner's handsome face, which is now ruined. One author sees a similarity here with Oscar Wilde's *The Picture of Dorian Gray* (Lachtman 1985, 134).

Another time we hear of Holmes using acid is in IDEN. Watson remarks upon returning to the Baker Street rooms that the odor of hydrochloric acid, HCl, told him that Holmes had spent the day working with his chemicals. Of course, there are the well-known stains that Watson reports. They are discussed in the concluding section of the chapter.

PHOSPHORUS

"Its muzzle and hackles and dewlap were outlined in flickering flame." Watson's description of the hound of the Baskervilles reflects the fact that Stapleton had applied some chemical to make the animal even more fearsome than his enormous size would warrant. The dog chasing Sir Henry Baskerville was "not a pure bloodhound and not a pure mastiff; but it appeared to be a combination of the two—gaunt, savage, and as large as a small lioness" (see figure 4.6).

Watson concludes that phosphorus was put around the mouth of the hound to produce the fearsome glow. That element, when exposed to air, glows in the dark. Phosphorus takes its name from the Greek for "light-bringing" (Greenwood and Earnshaw 1984, 546). Yet Holmes seems dubious that phosphorus was used. He notes that there is no odor from the chemical, so that nothing would interfere with the hound's sense of smell. Therefore, says Holmes, it must be, "a cunning preparation of it." Phosphorus produces its glow by reacting with oxygen in the air. When light emission is

Figure 4.6 The fearsome hound of the Baskervilles

produced by a chemical reaction, the process is called chemiluminescence, not phosphorescence.

Should we believe the physician who claims it is phosphorus or the chemist who is doubtful? When it comes to a chemical phenomenon, Holmes the chemist is probably a more reliable source than Watson the physician. It most likely was not phosphorus. What dog would stand to have phosphorus rubbed around its mouth? Sherlockians have suggested other materials that Stapleton could possibly have used rather than the unlikely phosphorus. For example, Redmond suggests barium sulfide, BaS (1964, 150). Whatever the chemical was on the hound's muzzle, it resulted in a terrifying appearance and caused Holmes to not hesitate before firing five shots to kill the mammoth dog.

AMALGAMS

If one would trust the chemist on glowing materials like phosphorous, then one might also believe he knows about amalgams. But it is not clear that Holmes does know amalgams. In ENGR, some "coiners" are making counterfeit coins. Holmes talks about an amalgam that they've used to replace the silver used in genuine coins. This is yet another instance where Asimov points out Holmes's error (Asimov 1980, 14). An amalgam is an alloy of mercury (Hg) and any other metal. Since Hg is an unlikely metal to be used in making coins, even counterfeit ones, Holmes may have misspoken.

In today's usage, "amalgamation" has come to mean any combination of things or even ideas. Perhaps Holmes was using the word in that sense. That does seem unlikely because he says that the amalgam has been used to replace silver in the coins, thus making an inexpensive substitute. On the other hand, when the counterfeiters flee, they leave behind "large masses of nickel (Ni) and tin (Sn)." These two elements, especially Ni, have been used in coinage. But the crooks leave no Hg. So were they using mercury at all? The absence of mercury in the house does make it possible that Holmes was using the mixture meaning of amalgam rather than the chemical term for "any alloy of mercury." I think, though, that the verdict on this issue must be that of Asimov: Holmes the chemist has made an error.

It is interesting to note that one of the counterfeiter's metals, Ni, is purified by use of one of the "poisons" discussed in section 4.4, carbon monoxide, CO. When impure Ni is reacted with CO (at 50°C), the metal carbonyl compound $Ni(CO)_4$ is formed. It is a toxic gas that can be collected, thus pulling the Ni metal away from the impurities. Then $Ni(CO)_4$ is heated to 230°C, which breaks it back down to metal and CO yielding 99.95 percent pure Ni metal:

$$Ni(CO)_4 = Ni + 4CO$$

This procedure for purifying nickel was developed in 1899 by L. Mond and is called the Mond process (Greenwood and Earnshaw 1984, 1330).

Section 4.6 Conclusion: Profound or Eccentric?

> *I gave my mind a thorough rest by plunging into a chemical analysis.*
>
> —Sherlock Holmes, *The Sign of the Four*

Anyone who finds a chemical analysis restful is clearly devoted to the science. There is no doubt that Sherlock Holmes loved his chemistry. He often became so engrossed in his experiments that he worked late into the night. Here are the words of Watson in SIGN:

> Up to the small hours of the morning, I could hear the clinking of his test-tubes.

And in COPP:

> Holmes was settling down to one of those all night researches which he frequently indulged in, when I would leave him stooping over a retort and a test-tube at night and find him in the same position when I came down to breakfast in the morning.

These odd hours are reminiscent of a university researcher (Gillard 1976, 10). Even his detective work sometimes was put aside so he could do chemistry:

> If there is an afternoon train to town, Watson, I think we should do well to take it, as I have chemical analysis of some interest to finish.

Holmes and Watson proceed to leave Norfolk in the middle of a case so that the chemistry analysis can be completed.

Holmes was dedicated to his chemical work, but what is the verdict on his chemistry skills? There are certainly indications of knowledge and skill. Graham (1945) has divided his chemistry efforts into two groups depending on whether they were related to crime detection or not. We've already looked at his "pure" (i.e., unapplied) chemistry earlier in this chapter. Rather than being the focus of a case, his forensic chemistry was sometimes merely mentioned by his chronicler Dr. Watson. In *Shoscombe Old Place* (SHOS), Holmes remarks on a case never narrated by Watson:

> Since I ran down that coiner by the zinc and copper filings in the seam of his cuff they have begun to realize the importance of the microscope.

Holmes has convinced Scotland Yard that the microscope is a useful investigative tool. He shows Watson the next objects he is examining by microscope—tweed threads, dust, epithelial cells, and glue:

> "Is it one of your cases?"
> "No; my friend, Merivale of the Yard, asked me to look into the case."

Holmes is assisting Merivale in the "St. Pancras" case. Sherlockians have debated whether Holmes could positively identify glue in this way. But the main point is that Holmes, and Scotland Yard, are beginning to use the microscope to identify chemicals. Given that the first use of the microscope in chemistry occurred in the 1700s, and that several books on the subject were published in the 1860s (Welcher 1957), it was high time for both Holmes and Scotland Yard to be using microscopes.

So we see Holmes as a devoted chemist working long hours on chemical analyses, interested in research into coal-tar derivatives, and capable of devising an important test for blood. In addition, we're told in STUD that Holmes had "extra-ordinary delicacy of touch" when manipulating his "fragile philosophical instruments." In FINA, as Holmes prepares for his confrontation with Professor Moriarty, he tells Watson of his plans to continue his chemical work in retirement:

> I could continue to live in the quiet fashion which is most congenial to me, and to concentrate my attention upon my chemical researches.

The Michells assert that Holmes was planning to continue his work on coal-tar chemistry. They even claim, without much justification, that he did so (Michell and Michell 1946, 250–251). Although Holmes displays many signs of a good chemist, there is also evidence of poor chemical technique in his work, and not just once. At their first meeting in STUD, Watson observes that Holmes's hands are "discoloured with strong acids." After sharing rooms with Holmes for awhile, Watson notes that his hands were invariably "stained with chemicals." Gillard (1976) notes "the cross-contamination of reagent bottles" that is described in NAVA, the twenty-fifth story. So Holmes has a long history of poor lab technique.

> He dipped into this bottle or that, drawing out a few drops of each with his glass pipette.

There is even a report that Watson initially described another instance of poor technique in RESI, but Conan Doyle deleted it from his final version (Cooper 1976, 70).

Holmes was delicate with his instruments, but dreadfully sloppy with chemical reagents. Watson also describes in SIGN one of his experiments that ended "in a smell which fairly drove me out of the apartment." In *The Empty House* (EMPT), the twenty-eighth story, Holmes's chemical working area at Baker Street is described as "acid-stained." By the time of the forty-ninth story, *The Mazarin Stone* (MAZA), it has become "acid-charred." Apparently Holmes continually spilled his chemicals.

But the most damning point against Holmes being a profound expert in chemistry is the fact that he lost interest in the subject long before his retirement. Holmes himself describes this change of interest in *The Abbey Grange* (ABBE):

> I propose to devote my declining years to the composition of a text-book, which shall focus on the whole art of detection in one volume.

So in FINA, the twenty-sixth story written in 1893, Holmes is planning that his retirement will be taken up with chemistry. By the time of ABBE, the thirty-ninth story written in 1904, he has changed his mind. Additionally, in *The Creeping Man* (CREE), the fifty-first story written in 1923, we hear directly about what interests Holmes now:

> He was a man of habits, narrow and concentrated habits, and I had become one of them. As an institution I was like the violin, the shag tobacco, the old black pipe, the index books.

There is no mention of chemistry. Holmes now looked elsewhere for diversion. He chose to tend his bees rather than work on his waning chemical interests. Ellison points out that, in the last two compilations of short stories totaling nineteen tales, there is no mention of Holmes doing chemistry (Ellison 1983, 36). In Tracy's encyclopedia, we find thirteen references to chemistry in the first thirty stories and only two references in the last thirty (1977). It is notable that the first half of the Canon is much more highly regarded than the second. When Holmes was depicted as a man of science, the stories worked much better than when science was absent.

It is entirely logical that Sherlock Holmes should move away from chemistry. His creator had done so. In his later life, Conan Doyle became one of the world's leading proponents of spiritualism, the belief that the spirits of the dead can communicate with the living. Generally debunked today, spiritualism enjoyed a period of wide acceptance. It began in America in 1848 in Hydesville, New York. There the young Fox sisters, Margaret and Kate, began the séances and levitations that they, forty years later, admitted were fraudulent (Miller 2008, 354). The movement took hold and claimed ten million American adherents by 1859. Its spread to England was aided by Queen Victoria attending séances there (Miller 2008, 353). In 1883, Conan Doyle wrote a story, *Selecting a Ghost,* which poked fun at the occult. But by 1885, he was attending sessions, though still harboring doubts. Hunting for a belief system to replace the rejected Catholicism, Conan Doyle examined telepathy, mesmerism, Buddhism, theosophy, and others (Miller 2008, 355). Gradually spiritualism gained ascendancy with him. In October 1917, Conan Doyle "crossed a Rubicon" (Lellenberg et al. 2007, 634) when he gave a public lecture that made clear his belief in spiritualism. Conan Doyle had a brief relationship with the world-famous magician Harry Houdini, a debunker of spiritualist phenomena. Both were desirous of convincing the other. Neither did, and it ended badly. Much of Conan Doyle's late life was spent writing about and lecturing on spiritualism.

Conan Doyle's non-Holmesian work included enough short stories to justify an anthology entitled *The Best Horror Stories of Arthur Conan Doyle* (McSherry et al. 1989). But we can be thankful that he kept things supernatural out of the Sherlock Holmes Canon. Toward the end, he did write a Holmes story entitled *The Sussex Vampire* (SUSS), the fifty-second of the sixty stories, published in 1924. But it has no vampire in it, and in discussing the possibility of vampires being involved in the case, Holmes tells Watson:

> Rubbish, Watson, rubbish! It's pure lunacy.

We conclude this chapter with an answer to the question twice posed: Is Holmes's chemistry profound or eccentric? Even after defending Holmes against most of Asimov's criticisms, I find that Watson's first opinion ("Knowledge of Chemistry—Profound") cannot be sustained. His blood test was not adopted by Britain. He tells us in EMPT that his work on coal-tar derivatives was finished to his satisfaction. That work may well have been his only chemical success. Had he remained interested in chemistry and had more success with it, his reputation might have warranted the adjective "profound." But his modest record requires that we rank Holmes the chemist somewhere between Watson's "profound" and Asimov's "blundering." "Eccentric" sounds just about right. After all, everything about Sherlock Holmes was eccentric.

5

Sherlock Holmes

Other Sciences

Section 5.1 Mathematics

It's a simple calculation enough.

—Sherlock Holmes, *A Study in Scarlet*

INTRODUCTION

Sherlock Holmes knew more chemistry than any other science. But in this chapter, we shall find that he was well informed in a number of other sciences as well. Since mathematics contributes to all sciences, we first examine the Canon for instances of mathematical knowledge. We find a number of references to and uses of math in nearly all the early stories. After Holmes and Moriarty supposedly went over the Reichenbach Falls in *The Final Problem* (FINA) and Holmes returned, he rarely used math again.

In *A Study in Scarlet* (STUD), Watson scoffs at a magazine article that claims that the conclusions of a trained observer are as "infallible as so many propositions of Euclid." He soon learns that his new roommate Holmes is the author of the article. So here, very early on, we have Holmes drawing a mathematical analogy to his deductive work. He invokes Euclid[1] again in the second story, *The Sign of the Four* (SIGN). This time he chides Watson about his writing style. Holmes accuses Watson of allowing romanticism to creep into his narration of the previous case, STUD. According to Holmes, this awkward technique produces "much the same effect as if you worked a love-story or an elopement into the fifth proposition of Euclid." The fifth proposition states that if two sides in a triangle are equal, then

[1] A Greek mathematician living around 300 B.C.

the angles opposite those two sides will also be equal. Note that Holmes makes no calculation using Euclid's proposition, but he depends on Watson's knowledge of math to make a point about the way the narrative of STUD was written. This is the first time, but not the last, that he criticizes Watson the chronicler.

In SIGN, Holmes's conversation again assumes that his listener is acquainted with mathematical terms. When he sees that Tonga has left a footprint in creosote, he claims that tracking him will be as easy as using the "rule of three," which states that if three of the four terms in a proportion are known, then the fourth may be calculated. It may be expressed as the following:

$$ad = bc \quad \textbf{or} \quad a{:}b{::}c{:}d \quad \textbf{or} \quad a/b = c/d$$

Knowing a, b, and c allows d to be calculated from $d = bc/a$. The equation can be arranged so that any of the terms can be computed. In nineteenth-century England, this rule had enough importance to be given a name. Today it is considered so mathematically trivial that one hardly ever hears of the "rule of three." Instead the operation performed is described as "the product of the means equals the product of the extremes," or cross multiplying. As with Euclid's proposition, Holmes does not use the rule of three to make a calculation in SIGN.

These mathematical references set a tone in the first two stories, STUD and SIGN. Here we have two learned men whose everyday conversation reflects a superior English education. Watson may later be befuddled by some of Holmes's deductions, but he is certainly no fool. Holmes uses mathematical terms as late as the fifty-seventh story, *The Lion's Mane* (LION), which is one of the two stories that he himself narrates. In LION, he describes the math teacher Ian Murdock as living "in some high abstract region of surds and conic sections." Conan Doyle certainly has a high opinion of his readers. He assumes they will know that a surd is a sum that contains one or more irrational roots of numbers.

HEIGHT FROM THE LENGTH OF A STRIDE

As noted in the opening quote of this chapter, Holmes does perform a calculation that he describes as simple. It is not considered simple today. So let's take a look at his determination of a suspect's height from the length of his stride. In STUD, Holmes examines the crime site where Enoch Drebber's body was found. He then gives Inspectors Lestrade and Gregson a number of clues. One of them is that the murderer is taller than six feet. Not only the Scotland Yarders are skeptical; Watson is too. Later he asks Holmes to explain how he deduced the man's height. Holmes replies:

> Why, the height of a man, in nine cases out of ten, can be told from the length of his stride.

This remark has been hotly debated in the Sherlockian literature. Many consider the calculation to be meaningless. They note that the length of a person's stride will vary with conditions. Yet even today, 125 years after Holmes was doing this calculation, it is not difficult to find online sites where a formula for the "simple" calculation is given. The following formulas for the calculation can be found at www.livestrong.com/article/438560-the-average-stride-length-in-running.

Height = 2.41(Stride) Males
Height = 2.42(Stride) Females

Even these sites admit variability and provide an alternate formula for a person running:

Height = 0.741(Stride) Athletes running

Campbell (1983, 15) gives a somewhat different formula: Height = 2.09(Stride).

Holmes again makes the claim to be able to use the length of Jonathon Small's stride to calculate his height in SIGN. In *The Boscombe Valley Mystery* (BOSC), Holmes claims the murderer is "a tall man." He tells Watson that it is a rough estimate from stride length. So already by the sixth story Holmes moderates his claim about the stride/height relationship. After mentioning it in three of the first six stories, Holmes never uses it again.

Today the U.S. Federal Bureau of Investigation doesn't use a stride/height relationship, believing it to be unreliable (Fisher 1995, 281):

> Contrary to the plotting of detective fiction, it isn't possible to estimate someone's height by the distance between steps—his gait—because during the commission of a crime, a suspect is usually moving very fast; he is running or backing up or moving sideways or struggling, attacking or defending, even sneaking around. The thing he isn't doing is moving normally.

Modern forensics attaches more relevance to foot size than to stride length in estimating a person's height (Ozden et al. 2005).

PROBABILITY

The only story from the second half of the Canon that contains any math is *The Six Napoleons* (SIXN), the thirty-fifth story. The tale involves six plaster of Paris busts of Napoleon. Someone is breaking into houses, stealing them, and then smashing them to bits. Such bizarre behavior leads to Watson's failed attempt at psychoanalysis (see section 2.2). He suggests that the culprit Beppo suffers from monomania. Here Holmes uses an elementary calculation of probabilities. When only two of the six busts remain, Holmes states that there is a two-thirds probability that the burglar will strike again. Why two-thirds?

Inspector Lestrade consults Holmes because the case is so "outré." Although there are hundreds of busts of Napoleon in London, the thief Beppo is interested only in the six that were made at the same time about a year ago. When the fourth bust is stolen from Horace Harker, a journalist, a dead body is found with it. At this point, Lestrade loses interest in the busts; he has a murder to solve:

> After all, that is nothing; petty larceny, six months at the most.
> It is the murder that we really are investigating.

Of course Holmes sees a connection and continues to focus his interest on the busts of Napoleon. Holmes mentions to Lestrade that all four stolen busts were immediately smashed where there was enough light to examine the pieces. Lestrade fails to see the significance of this and continues to seek information about the dead man. Holmes's knowledge of past crimes enables him to unravel this mystery. Recall Watson's earlier assessment:

> Knowledge of Sensational Literature—Immense.
> He appears to know every detail of every horror perpetrated in the century.

Holmes, the student of crime, again has an edge on the Scotland Yard official force. Only Holmes remembers that the theft of the black pearl of the Borgias had happened about a year ago, just as the busts were being cast at the firm of Gelder & Co. He deduces that the missing jewel is in one of the busts. Beppo, formerly employed at Gelder's, had hidden the stolen pearl in one of the Napoleon busts as it was being made. He did this while being pursued by police for knifing another man. After a year in prison, he wanted to regain the pearl.

Holmes's theory explains why the fourth bust was smashed under a street lamp. Lestrade doesn't care where the busts were broken. Having murdered a competitor at the fourth theft, Beppo must act swiftly. Thus Holmes concludes he will strike again the next night. In order to convince Beppo that the police are on the wrong track, Holmes tells the journalist Harker that he agrees with Lestrade's opinion that this is the doing of a Napoleon-hater. Harker publishes that idea in the newspaper.

Two busts remain. One is nearby in Chiswick. The other is thirty-five miles away in Reading. Holmes persuades Lestrade to accompany him to Chiswick the next night by proclaiming that there is a two-to-one chance they'll apprehend the thief/murderer. Since there are only two busts left, why are the odds two to one? Holmes knows that the pearl wasn't in any of the first three busts because a fourth one was stolen. But he doesn't know if Beppo had success with the fourth. Perhaps the pearl was in the fourth bust. If not, it must be in one of the last two. Thus there is one chance that Beppo has the pearl, and two chances that it remains encased in plaster, ergo Holmes's comment that the chances are two to one they'll make an arrest in Chiswick.

Holmes is pretty certain that if Beppo still doesn't have the pearl, he will strike the next night at Chiswick. He will not go to the distant Reading. Gauging the probability of an arrest to be two-thirds, Holmes coaxes Lestrade to join him in Chiswick. In making the statement that he has a two-thirds chance of nabbing Beppo, Holmes makes two very good assumptions. He presumes Beppo will strike the very next night. This is likely because, with murder now involved, the police will be expending more effort on the case. So Beppo is likely to make haste with his next attempt to recover the black pearl of the Borgias. Holmes also reasons that the nearby Chiswick will be Beppo's next target, not the distant Reading. Sherlock's reasoning proves correct, and Beppo is apprehended in Chiswick. Lestrade looks to be a real bungler in this story. He misses key points and goes off on a tangent.

GEOMETRY AND THE RULE OF THREE

In chapter 4, we saw that Holmes was able to follow the directions of the Musgrave ritual to find the small cellar room where the ancient crown of the king of England had been concealed. There, instead of the crown, he found the body of Brunton the butler who had followed the ritual before Holmes. Here we look at the geometric calculations required to follow the ritual.

The directions are simple to follow:

> "How was it stepped?"
> "North by ten and by ten, east by five and by five, south by two and by two, west by one and by one, and so under."

Holmes's problem was where to start. Both Holmes and Brunton conclude that the starting point was the tip of the shadow of the elm tree at a certain position of the sun. But the elm tree is gone, having been felled by lightning ten years earlier. But Reginald Musgrave knows that the elm was sixty-four feet high. His geometry tutor years earlier had him do a number of such calculations. To be able to use the elm's height to calculate the length of its shadow, Holmes erects a six-foot fishing pole at the site of the elm stump. Its shadow, when the sun was "over the oak" is nine feet high. This permits him to set up the proportion below and thereby compute the length of the elm's shadow to be ninety-six feet:

Shadow Pole/Ht. Pole = Shadow Elm/Ht. Elm
9/6 = Shadow Elm/64
Shadow Elm = 96

Unlike SIGN, Holmes actually uses the "rule of three" in this story to make a computation. He then locates his starting point at a distance of ninety-six feet from the stump in the same direction as the shadow of the pole. Following the commands of the Musgrave ritual, he then locates the cellar room where Brunton's body is found.

The entire calculation involving the elm tree is dependent on the height being the same as it was 250 years ago when the crown was hidden. Elm trees can grow higher than sixty-four feet. Climate and soil are important factors in any tree's mature height. Did this tree stay at that height for 250 years? As we've seen, Holmes is very much aware that the length of his stride depends on his height. He would also know that humans were shorter when the crown was hidden over 200 years ago. So, although we are not told that he adjusts his stride as he follows the paces described in the ritual, we can be confident that he did.

MENTAL MATH

In section 3.7, we noted that *Silver Blaze* (SILV) contains the most famous words written in all of the sixty Sherlock Holmes stories: "the dog did nothing in the nighttime."[2] This is the famous "enigmatic clue." Holmes makes another notable statement in SILV, a mathematical one. At the beginning of this adventure, when Holmes and Watson take the train to Dartmoor, he remarks that

> [o]ur rate at present is fifty-three and a half miles an hour.

Then, by way of explanation, Holmes says,

> The telegraph posts upon this line are sixty yards apart, and the calculation is a simple one.

[2] There are about 800,000 words in the Canon (Swift and Swift 1999, 37).

Figure 5.1 Sherlock Holmes checking his watch

At first glance, the reader will not see an easy connection from sixty-yard gaps between posts to fifty-three and a half miles per hour. How did Holmes do this calculation in his head? His watch was the only device he used (see figure 5.1).

Sherlockians have proposed several methods for this mental math. All of them start by constructing an equation with three unknown quantities: the time, the number of gaps traversed in that time, and the train's speed. Holmes then measures the time and number of gaps, allowing him to compute the speed. But can it be done in such a way that the calculation is simple? Here is Bengtsson's explanation (Bengtsson 1989).

Let's begin our analysis by establishing the equation that relates our three quantities. First, speed is distance divided by time:

$$S = D/t$$

The total distance traveled will be N gaps times 60 yards per gap:

$$D\ (\text{yards}) = 60N$$

To convert this to miles, we divide it by 1,760 yards per mile:

$$D\ (\text{miles}) = 60N/1760$$

To convert this to a speed, we must divide distance by time, t(sec):

$$S(\text{miles/second}) = D/t = 60N/1760t$$

To convert this to hours, we must multiply by 60 sec/min and also by 60 min/hour:

$$S(\text{miles/hour}) = [60N/1760][60 \bullet 60/t] = (N/t)[60 \bullet 60 \bullet 60/1760]$$
$$= (N/t)[6 \bullet 60 \bullet 60/176]$$

It still isn't the least bit simple, until Holmes has the mathematical insight to note that $176 = 11 \bullet 16$:

$$S(\text{miles/hour}) = (N/t)[6 \bullet 60 \bullet 60/176] = (N/t)[6 \bullet 60 \bullet 60/11 \bullet 16]$$

Then it gets easy. He sees that making N = 11 will eliminate that awkward factor from the equation. So he does that by measuring the time to travel 11 gaps:

$$S = (11/t)[6 \bullet 60 \bullet 60/16 \bullet 11] = [6 \bullet 60 \bullet 60/16t] = [6 \bullet 60 \bullet 60/4 \bullet 4 \bullet t]$$
$$[6 \bullet 15 \bullet 15/t]$$
$$S(\text{miles/hour}) = 1350/t$$

It turned out that the train's speed was such that as they approached the twelfth pole (11 gaps), Holmes saw that the time was nearing 25 seconds. It is relatively simple to see that 1,350/25 = 54 (four 25s in a hundred, thus fifty-two 25s in 1,300. Then add two more 25s for the 50 bringing the number of 25s in 1,350 to 54). That meant that if the train reached the twelfth pole in exactly 25 seconds, the speed would have been 54 mph. When the train didn't quite reach the last pole in 25 seconds, Holmes merely gave a good estimate that somewhat less than 54 could be described nicely as the 53.5 mph that he reported to Watson. Some agonizing discussions exist claiming that the number of significant figures he gave, three, denotes a more exact calculation than the one just described. Today's students, who often report as many figures as their calculators can give, will realize that in this setting (i.e., not a research lab), it is not an important issue. The main point is that Holmes's remark that the speed is 53.5 mph illustrates his facility with mental calculations.

Section 5.2 Biology

> *Which is it today, morphine or cocaine?*
>
> —Dr. Watson, *The Sign of the Four*

ANATOMY

In his rating of Holmes's abilities in STUD, Watson rates Holmes separately in two areas of biology, botany and anatomy. As was the case with mathematics, more than two-thirds of the biological references occur in the first half of the Canon. According to Watson, Holmes's knowledge of anatomy is "accurate, but unsystematic." This probably refers to the fact that, as usual, Holmes has learned only what anatomy he felt could help him as a consulting detective, "the only one in the world." We learn in the second story, SIGN, that Holmes has already authored a monograph with a rather lengthy title:

> The Influence of Trade Upon the Form of the Hand, With Lithotypes of the Hands of Slaters, Sailors, Cork-cutters, Compositors, Weavers, and Diamond-polishers

Holmes actually uses this ability in some cases. In *A Case of Identity* (IDEN), his first words to Mary Sutherland are:

> Do you not find that with your short sight it is a little trying to do so much typewriting?

He later explains to Watson that he could see an impression above her wrist where it rested on the typewriter. This type of observation is reminiscent of Dr. Joseph Bell. Recall from section 1.4 when Bell deduces that a woman is a linoleum worker from the dermatitis on the fingers of her right hand. The fact that Mary Sutherland is a typist does play a role in the plot, although it does not help Holmes arrive at a solution to the case. IDEN is the story that Holmes solves using the idiosyncrasies of a typewriter. The culprit is Mary's stepfather, James Windibank (see section 3.5)

In *The Solitary Cyclist* (SOLI), Holmes looks at the anatomical features of Violet Smith's hands and deduces that she is a musician. He admits he almost thought that she too was a typist because musicians and typists have similar hand types. But he eventually got it right. The problem in SOLI is that Violet Smith is not a solitary cyclist. She is being followed by another cyclist. This alarms her, and she consults Sherlock Holmes. Again the fact that Holmes could deduce her profession plays no role in solving the case. Holmes sends Watson to investigate the second cyclist. When he receives Watson's report, the unsympathetic Holmes tells the good doctor,

> You really have done remarkably badly.

But Holmes is able to stop the plans of Jack Woodley and Bob Carruthers to get at Violet's fortune by forcing her to marry Woodley.

Another part of human anatomy that Holmes studied was the finger. He saw the potential use of fingerprinting in crime-solving before Scotland Yard did. In chapter 3, we discussed several stories where fingerprints are mentioned, with the only significant usage occurring in *The Norwood Builder* (NORW) when John Hector MacFarland's right thumbprint is found on the wall.

A third part of human anatomy that drew Holmes's interest is the ear. In *The Cardboard Box* (CARD), Holmes claims to have authored two monographs on ears in the *Anthropological Journal*. He believes that "[e]ach ear is as a rule quite distinctive and differs from all other ones."[3]

It is an unusual detective story in which the shape of an ear plays a significant role. But that is exactly the case in CARD. Susan Cushing receives a cardboard box through the mail. In it are two severed ears, one a woman's ear, the other a man's. Holmes's solution is hastened when he notices the strong ear resemblance between the severed female ear and that of Susan Cushing. Soon he has shown that the murderer is Jim Browner, husband of Susan Cushing's youngest sister Mary. In a fit of rage, Browner has killed his wife and her lover Alec Fairbairn. He then sends the severed ears to the third sister, Sarah Cushing. It was Sarah who coveted Jim and then sabotaged his marriage when he rejected her. But Susan Cushing mistakenly receives the ears, consults Holmes, and justice is done.

The hamstrings, the muscles at the back of the thigh, are mentioned in two stories. In *The Musgrave Ritual* (MUSG), the body of Brunton the butler is found "squatted down upon his hams." The hamstrings play a more significant role in SILV. John Straker was killed by Silver Blaze while attempting to "make a slight nick upon the tendons of a horse's ham" (i.e., the hamstrings). Holmes is delighted that his inquiry about sheep reveals that three had recently gone lame. His reasoning was that the culprit would want to practice his tendon-snipping skill. And why would the trainer sabotage his own horse's chances? Straker

[3] Are they?

planned to bet heavily on the opposition horse. He wanted Silver Blaze to be able to run, just not too fast.

In the sixtieth story published, *Shoscombe Old Place* (SHOS), Sir Robert Norberton has concealed his sister's death by placing her body in the church crypt in a coffin previously occupied by an ancestor. He has some of the ancient bones burned at night. But one of the stable lads finds an old femur before it is burned. He takes it to John Mason, the head trainer of Norberton's horse Shoscombe Prince. Mason consults Sherlock Holmes, who asks Dr. Watson:

> "What do you make of it, Watson?"
>
> "It's the upper condyle of a human femur."
>
> "Exactly!"

Tracy defines a condyle as "a protuberance on the end of a bone serving to form an articulation with another bone" (1977, 82). The human femur has just such a thing at the lower end of it (i.e., the knee). But there is no such thing as an upper condyle on a human femur, where it joins the hip. Note that it is Watson who makes the mistake first. But Holmes enthusiastically agrees. We could let them the share the blame for the scientific error. Or perhaps Conan Doyle deserves the blame.

Norberton's motive is his desire to avoid bankruptcy. He must prevent news of his sister's death from reaching his creditors until Shoscombe Prince wins the derby. The condyle, whether it is upper or lower, is just a minor clue to the proceedings. The behavior of the spaniel is much more important (see section 3.7).

We talk more about the condyle in "Doyle Scams" in the appendix.

BOTANY

Watson, in STUD, rates Holmes's botany as "variable." Furthermore, Holmes is

> well up in belladonna, opium, and poisons generally.
> Knows nothing of practical gardening.

We first examine Watson's opinion of Holmes the gardener. One instance of Holmes displaying his mediocre botanical skills is his behavior in *Wisteria Lodge* (WIST). The case takes Holmes to the village of Esher in Surrey. Holmes needs some kind of cover while he keeps a nearby house under surveillance. He tries to divert attention by reading an elementary book on botany and collecting botanical samples while he keeps watch on the house. But, according to Watson, "it was a poor show of plants which he would bring back of an evening." Watson's assessment of Holmes and gardening is accurate.

Because they directly affected his work, Holmes had much greater interest in poisons. We looked at chemical poisons in chapter 4. There is also frequent mention of biologically based poisons in the Canon. Such substances began to replace inorganic poisons around the middle of the nineteenth century. These molecules were discovered or isolated starting with morphine, a derivative of opium, in 1804. Others soon followed: nicotine (1807), strychnine (1819), and cocaine (1860). When the Marsh test and then the Reinsch test of 1842 (Wagner 2006, 51) made detection of arsenic reliable, poisoners looked to abandon this "inheritance powder." They started to use biological poisons with greater frequency. In this way they were able to stay ahead of law enforcement's detection capabilities. In the middle of the nineteenth century, authorities in France ruefully reported that (Blum 2010, 2)

> [h]enceforth let us tell would-be poisoners; do not use metallic poisons for they leave traces. Use plant poisons. Fear nothing; your crime will go unpunished.

By 1851, when "the potent poison" nicotine (Wagner 2006, 56) was first detected in a corpse, bio-poisons were being used regularly in murders. Interestingly one such case was solved by Dr. Henry Littlejohn (see section 1.3) in 1878. He was able to get a conviction by showing that opium was the cause of death (Wagner 2006, 55). The Holmes stories start off with biological poisons being used for murder in the first two stories. In STUD, the Mormon Enoch Drebber is killed by what is probably curare. Jefferson Hope gets his revenge on Drebber by using an alkaloid extracted from a South American arrow poison. Curare is the most famous of these arrow substances (Tracy 1977, 94). In SIGN, Tonga kills Bartholomew Sholto using a "strychnine-like" substance, resulting in a very unpleasant death (Cooper 2008, 41). Presumably the dart from Tonga's blow gun that just misses Holmes and Watson during the concluding high-speed boat chase down the Thames has the same deadly compound on the tip.

Bio-poisons were used in several other Holmes cases, without causing death. In *The Sussex Vampire* (SUSS), the jealous Jack Ferguson fails in his effort to kill his young half-brother with curare. In SILV, the stable boy Ned Hunter is drugged with powdered opium. Conan Doyle's description of the tranquilizing effect of opium is accurate. With Ned Hunter in somewhat of a stupor, John Straker is able to remain undetected as he leads Silver Blaze out onto the nearby moor. Then as Straker attempts to snip a tendon, the frightened horse rears up and strikes the trainer with his hoof, killing him.

Figure 5.2 His eyes rested thoughtfully upon the sinewy forearm and wrist

Sherlock Holmes's cocaine issues were introduced in chapter 2. It has been pointed out that Conan Doyle's description of Holmes's reaction to drugs doesn't match reality. He describes cocaine as a tranquilizing drug when it actually tends to stimulate (Pratte 1992). In several stories, we see Holmes's need for mental stimulation. In *The Hound of the Baskervilles* (HOUN), he tells Watson that his afternoon was spent consuming "two large pots of coffee and an incredible amount of tobacco." We saw earlier that, in *The Red-Headed League* (REDH), the case caused him to turn to nicotine to help him solve a "three pipe problem." In *The Missing Three-Quarter* (MISS), he complains of stagnant days (see figure 5.2).

In SIGN, after injecting himself with the famous 7 percent solution, Holmes says that his mind rebels at stagnation:

> I crave mental exaltation.

Watson's emotional response:

> Count the cost!
> Why should you, for a mere passing pleasure, risk the loss of those great powers with which you have been endowed?

Watson talks in MISS of how he has weaned Holmes from the drug habit "which had threatened once to check his remarkable career." But even as he thinks he has gotten Holmes's off the cocaine, he tells us,

> I was well aware that the fiend was not dead but sleeping.

The treatment of poisons and drugs in the stories is, of course, shaped by Conan Doyle's attitude toward them. On September 20, 1879, he wrote a letter to the *British Medical Journal* entitled "Gelseminum as a Poison" (Gibson and Green 1986, 13). In order to test the poisonous properties of gelseminum, he administered a small amount to himself. He kept increasing the amount everyday until he could no longer stand it:

> The diarrhea was so persistent and prostrating, that I must stop at 200 minims. I felt great depression and a severe frontal headache.

This same concept appears immediately in the very first Holmes story, STUD. In the opening chapter, Young Stamford warns Watson:

> I could imagine his giving a friend a little pinch of the latest vegetable alkaloid, not out of malevolence, you understand, but simply out of a spirit of inquiry.

Stamford adds that Holmes would also readily take some alkaloid himself in order to learn about its effects.

The most interesting aspect of the Holmes/cocaine scene is how Conan Doyle has Watson condemn its use. SIGN, in which Watson tells Holmes to count the cost, was published in 1890. The prevailing view of cocaine was rather positive at that time. In 1884, Sigmund Freud wrote a review article on cocaine that he described as "a song of praise to this magical substance" (Musto 1968, 128). He tells us about experimenting on himself with cocaine. Perhaps Freud helped to influence Conan Doyle's view of testing substances on oneself.

Freud's final article on cocaine was published in 1887. It still speaks positively about cocaine, but somewhat less so. In it, Freud cites supporting statements by William A. Hammond (Musto 1968, 129), the U.S. Surgeon General during the Civil War. Following the war, he was a very successful physician in New York City (Sartain 2008). He announced that cocaine was a harmless tonic that cheered the melancholy while having no adverse side effects, and that it was not addictive (Musto 1988, 215). Hammond felt that a cocaine habit was very much like a coffee habit (Musto 1968, 130). Indeed, Musto tells us that "[c]ocaine as Holmes used it was in accord with the advice of leading physicians." Despite all the praise from these two respected authorities, Conan Doyle was early in his negative appraisal of the effects of cocaine. Note that, contrary to Hammond's view, he describes it as addictive. Watson has to wean Holmes from it, and still feared that the "fiend" would return. In this instance, Conan Doyle the physician was ahead of his time. Cocaine would become universally condemned thereafter.

Section 5.3 Physics

He threw himself down upon his face with his lens in his hand.

—*The Speckled Band*

OPTICS

The public often connects Sherlock Holmes with a magnifying lens. This is no wonder because it has been called "the very first tool of deduction" (Capuzzo 2010, 14). It is one of several optical devices that are used in the Holmes stories, mentioned in twenty of the sixty (Coppola 1995, 110). In the first story, STUD, Holmes spends twenty minutes examining the room where the body of Enoch Drebber was found. Watson describes him as "sometimes stopping, occasionally kneeling, and once lying flat upon his face." In the next adventure, SIGN, Holmes makes even more use of his lens. His deductions about Watson's brother's watch (see section 2.1) follow his examination of it with a convex lens. Holmes next uses his lens to examine the rope that Jonathon Small used to climb into Bartholomew Sholto's room. His third use of a lens in SIGN is to study the room where the murder was committed:

> He whipped out his lens and a tape measure and hurried about the room on his knees, measuring, comparing, examining with his long thin nose only a few inches from the planks.

Isn't that the Holmes we love, so intent when he is hot upon the trail? Here is Watson's description in BOSC:

> His face flushed and darkened. His brows were drawn into two hard black lines, while his eyes shone from beneath them with a steely glitter.

> His face was bent downward, his shoulders bowed, his lips compressed, and the veins stood out like a whipcord in his long sinewy neck. His nostrils seemed to dilate with a purely animal lust for the chase.

In BOSC, Holmes uses his lens to examine the ground around Boscombe Pool. This examination leads to the solution of the mystery. In REDH, he uses his lens to examine the cracks between the stones in the floor through which the bank robbers will dig their way into the vault. Somehow this enables him to predict that it will be another hour before they climb up to their capture.

We've already seen how Holmes was able to make accurate deductions about Dr. Mortimer in HOUN and Henry Baker in *The Blue Carbuncle* (BLUE). In

Figure 5.3 Holmes used a magnifying glass throughout his career

those cases, he used his magnifying lens on Mortimer's walking stick and Baker's hat (see figure 5.3).

In NORW, Holmes examines the vital thumbprint clue with his lens. In *The Beryl Coronet* (BERY), it is footprints on the windowsill that he magnifies for examination. This allows him to trace the movements of the suspects and show that Arthur Holder did not steal the priceless coronet. The lens is again used to examine the footprint of a shoe on a windowsill in *The Valley of Fear* (VALL). In *The Golden Pince-Nez* (GOLD), Holmes looks through his lens at a telltale fresh scratch mark on the lock of Professor Coram's bureau. It's an important clue. The lens is used in several other cases as well, though without large impacts. He examines a bush in *The Bruce-Partington Plans* (BRUC), the crucial gash in the stone in *The Problem of Thor Bridge* (THOR), a blood mark on a notebook in *Black Peter* (BLAC), and the lamp in *The Devil's Foot* (DEVI).

It is notable that Holmes continued to use his magnifying lens throughout the Canon. After a flurry of early use in six of the first thirteen stories, he uses it nine more times in the last forty. These nine times are spread evenly throughout the stories. He may have drifted away from chemistry, biology, and math, but he stuck with his lens. It is generally true that his most effective uses of it were in the early stories. However, he did well with the lens in story number fifty-seven, *The Lion's Mane* (LION). By this time, Holmes has left London and is living in a villa on the south coast "commanding a great view of the Channel." Unsurprisingly he gets involved in local events. For example, just before the local science teacher Fitzroy McPherson dies, he utters the words "the lion's mane." Holmes is confused by these words, as well as the strange marks on the body.

Holmes uses his lens to examine McPherson's body. He then reminds us, for LION is one of the two cases narrated by Holmes instead of Watson, "that I hold a vast store of out-of-the-way knowledge." He finally remembers a book that describes wounds such as McPherson's and attributes them to *Cyanea capillata*, a type of jellyfish also called the Lion's Mane. Inspector Bardle of the Sussex Constabulary, anxious to make an arrest for murder, asks for Holmes's help. Thanks to the lens and his vast store of knowledge, Holmes is able to show that no murder has been committed.

Two other optical devices get a mention in the Holmes stories. A telescope is used in HOUN. Mr. Frankland uses it to keep tabs on all that happens on the moor (see figure 5.4). When Watson looks through the telescope, he too sees suspicious activity and immediately goes out onto the moor to investigate. He is shocked to find that the mysterious person living out there is none other than Sherlock Holmes. Holmes and Watson had been working separately on the Baskerville case, but from now until the end of the case,

Figure 5.4 Frankland and Watson use a telescope to watch the moor in *The Hound of the Baskervilles*

they work together. That is the only role that a telescope plays in the Canon. There is not even an entry under "telescope" in the Holmesian encyclopedias (Tracy 1977; Bunson 1994; Park 1994).

We saw in chapter 4 that Holmes uses a microscope in SHOS. Though his work with it was successful in SHOS, the case was one we never hear about, as it was something he was working on simultaneously with the events surrounding Shoscombe Prince. Watson tells us little else about Holmes and the microscope. Because SHOS is the very last story published, we can take Holmes's use of the microscope as evidence that he was evolving as a forensic detective. Though he never gave up the magnifying glass, he was looking to the future by beginning to use the microscope as well.

OTHER PHYSICS

There are a few other aspects of Holmes's work that fall under the mantle of physics. One is his knowledge and use of gunshot residues, still important in

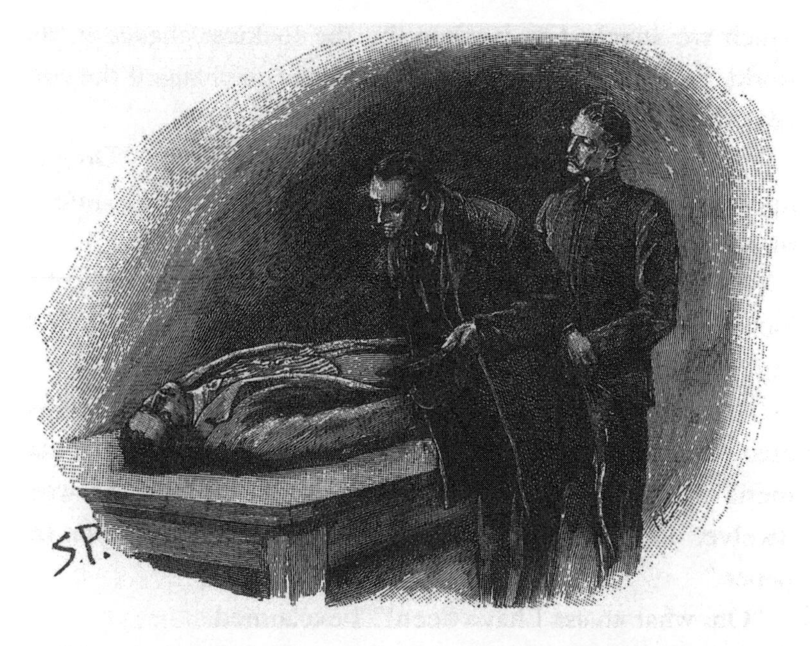

Figure 5.5 Holmes notes the absence of gunpowder on the corpse in *The Reigate Squires*

today's courtrooms. Holmes's first significant use of gunshot residues takes place in *The Reigate Squires* (REIG). William Kirwan, the coachman for the Cunninghams, is found dead. The Cunninghams, both father and son, claim to have seen the murderer. Young Alec Cunningham reports that Kirwan and his assailant were locked in a struggle when the fatal shot was fired. The murderer then fled. Upon examining the body, Holmes instantly concludes that Alec Cunningham is lying. The basis for this conclusion is that there is no powder mark on the dead man (see figure 5.5). As Holmes explains his reasoning at the end of the case, he remarks that the lack of powder had convinced him that the shot had been fired from a distance of more than four yards. Coupling this evidence with his brilliant deductions based on the handwritten note (see section 3.4), Holmes is able to make his case against the actual murderers, the Cunninghams.

In *The Dancing Men* (DANC), we get a somewhat different use of powder following a gunshot. In this story, it appears that Elsie Cubitt shot and killed her husband Hilton. She then failed in her attempt to kill herself. Holmes immediately rejects this official version. He had been contacted by Hilton Cubitt about messages being left at his house in the strange form of dancing men figures. Holmes has already cracked the code (see section 3.4) and knows there is another person involved in the case. Two servants, Saunders the housemaid and Mrs. King the cook, find the Cubitts, one dead and the other nearly so. They report that they could immediately smell gunpowder

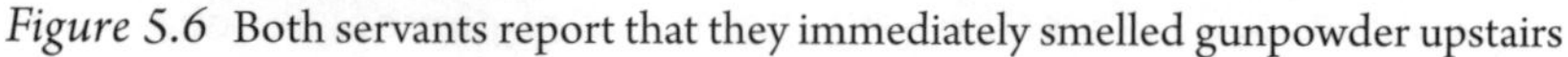

Figure 5.6 Both servants report that they immediately smelled gunpowder upstairs

upon hearing the shots and before exiting their upstairs rooms to come down to the study where the crime was committed (see figure 5.6). Holmes deduces that both the window and the door to the study had been open. Here he is applying knowledge of Graham's Law of Diffusion to the movement of the vapor through the house.

Thomas Graham, a Scotsman, formulated two laws that describe how fast gases move. His Law of Effusion can calculate how rapidly a gas will escape through a pinhole into a vacuum. Graham's Law of Diffusion deals with the speed with which two gases mix. That is the situation in DANC. Diffusion is more complicated than effusion and thereby more approximate. The gunpowder vapor in DANC must move through the air and arrive in the upstairs hall almost immediately. Holmes knows that couldn't happen unless there was the assistance of a breeze. This is a commonsense conclusion, but Inspector Martin is a bit slow to grasp it:

> "You remember, Inspector Martin, when the servants said that on leaving their room they were at once conscious of the smell of powder, I remarked that the point was an extremely important one?"
>
> "Yes sir, but I confess I did not quite follow you."

Holmes goes on to describe the murder scene. A third person was outside the window. He and Hilton Cubitt fired almost simultaneously, giving rise to the very loud noise that woke the staff upstairs. Cubitt was killed by Abe Slaney, but his bullet missed Slaney. Holmes looks for and finds evidence of a third bullet on the windowsill. Elsie Cubitt, distraught over her husband's death, then shot herself in the head.

In BRUC, Watson tells Holmes,

> A masterpiece. You have never risen to a greater height.

What has impressed Watson is mainly physics, again mixed with commonsense. BRUC is one of the cases involving Mycroft Holmes. The plans for a submarine have gone missing. Worse than that, Arthur Cadogan West is dead. His body is found next to the London Underground train tracks near the Aldgate station. BRUC was published in December 1908. Another dead body was also found at the Aldgate station in *The Mysterious Death on the Underground Railway* (Ackroyd 2011, 168), also published in 1908. This novel was perhaps one of Conan Doyle's sources.

In BRUC, no ticket for the train is found on the body. Only Holmes considers that important. He says to Inspector Lestrade,

> Why had he no ticket?

Lestrade's idea is that the murderer removed the dead man's train ticket before he tossed the body out of the train. Holmes's insight about the lack of a train ticket helps him deduce that Cadogan West was never in the train, but rather on top of it. He was killed in a flat on the edge of the tracks in one of the few areas where the Underground emerges, near the Gloucester Road Station. The murderer merely tossed the body onto the top of the train as it stopped nearby. The murderer may even have thought of this strategy after hearing guards along the Underground calling out (Ackroyd 2011, 140),

> It is forbidden to ride on the roof.

The body remained there, following the principles of friction, inertia, and momentum.[4] Friction had to be overcome before it could move. This occurred near Aldgate because of two factors: "Points, and a curve," says Holmes.

Aldgate is a junction, and the tracks curve. At a junction, the train goes over "points," making for a bumpy ride. The lessening of friction coupled with the

[4] A map of the 1908 London Underground suggests that Cadogan West's body remained on the roof of the Underground train for approximately twelve stops.

momentum as the train went round a curve flung the body over the side. Had the train gone smoothly in a straight line, the body would have remained on the roof.[5]

The last application of physics that we shall discuss has been a source of controversy in the Sherlockian literature for years. In *The Priory School* (PRIO), Holmes claims to know the direction in which a bicycle was traveling by examining its wheel tracks in soft ground. When he remarks to Watson that the bike was heading away from the Priory School, Watson responds,

> Or towards it?

Holmes answers,

> No, no, my dear Watson. The more deeply sunk impression is, of course, the hind wheel upon which the weight rests. You perceive several places where it has passed across and obliterated the more shallow mark of the front one.

In this story, Lord Saltire, ten-year-old son and heir of the Duke of Holdernesse, has been abducted from his school. Heidegger, who taught German at the Priory School, rode his bicycle desperately after Lord Saltire in an attempt to help him get away from his abductor. James Wilder, secretary to the duke and his illegitimate son, arranged the abduction by Reuben Hayes, because he wished to become heir to the duke's fortune.

When Holmes comes upon a set of bicycle tracks made by Dunlop tires, he makes his claim about the direction of the bicycle. His claim was immediately challenged by readers who thought that obliterated tracks could not enable direction to be determined. Holmesian scholars lined up on both sides of the debate (Baring-Gould 1967, vol. 2, 617). Conan Doyle soon heard that the point was under dispute (Haining 1995, 161):

> I dare say I have had twenty letters upon the one point alone.

He decided to use a bicycle and test the idea. He found that he could not tell direction on flat ground, but that he could on a hill. Actually, in the story, Conan Doyle had already hit upon the solution to the direction question. At one point Holmes says to Watson,

[5] An extensive analysis of the Underground and its motions was the subject of early Holmesian research (Crump 1952).

> Do you observe that the rider is now undoubtedly forcing the pace?

Holmes has observed that the tracks of both front and back wheels are equally deep:

> That can only mean that the rider is throwing his weight on to the handle-bar, as a man does when he is sprinting.

This principle, accepted by both sides of the debate, provides the answer to the direction question. Watson describes the terrain as "rolling hills." Conan Doyle found that going uphill resulted in deeper tracks by both wheels than going downhill. So Holmes could tell the direction of travel, but not by obliteration of the tracks.

There are two other notable features in PRIO. The first deals with the fact that Holmes can easily tell Heidegger's bike tracks from Wilder's. Heidegger's bicycle has Palmer tires and Wilder's has Dunlops:

> I am familiar with forty-two different impressions left by tyres.

Holmes's feat may have been less impressive than it seems. At the time, it was common for bike tires to bear the company logo on the tread (Klinger 2005, vol. 2, 948). When Heidegger's Palmer tracks are encountered, they lead to his body. Hayes has killed him with a blow to the head.

Holmes's claim about tire treads also brings to mind similar claims he made elsewhere. In IDEN, Holmes asserts that James Windibank's typewriter had sixteen different characteristics that were unique. In REIG, he says that there were twenty-three characteristics of the handwriting that would link the incriminating note to the Cunninghams. In HOUN, Holmes remarks that there are seventy-five perfumes that a criminal expert should be able to distinguish. In BOSC, he even claims to have authored a monograph on 140 varieties of tobacco (Smith 2011, 49). All this knowledge was gathered before we ever meet Holmes. It might have been interesting to read about this time in Holmes's life. But I suppose stories about a would-be detective getting an education might not be as entertaining as Conan Doyle's sixty tales.

Another aspect of PRIO caused Holmes to say,

> The case deserves to be a classic.

How did Reuben Hayes get out on the moor, abduct Lord Saltire, kill Heidegger, and leave no tracks? The only tracks other than those of Heidegger's and

Wilder's bicycle tires are from cows. Meditating on this over lunch, Holmes says to Watson,

"Well, now, Watson, how many cows did you see on the moor?"

"I don't remember seeing any."

Watson too now wonders about the cow tracks. He asks,

"And what is your conclusion?"

"Only that it is a remarkable cow which walks, canters, and gallops."

Holmes has recalled the pattern of the hoof marks and correctly deduced that the tracks were those of a horse. Hayes put shoes on the horse that made prints that looked like cow tracks. He was attempting to avoid blame by concealing that he'd been on the moor. He wasn't expecting Sherlock Holmes as the investigator.

When Watson assesses Holmes's abilities in STUD and also when he recollects his ratings in FIVE, he says nothing about physics. Given the nature of the physics in the stories, this is understandable. Holmes's physics is mainly commonsense reasoning that many could do without knowing the physical principle being applied. That he does it better and quicker than the official police force shows again that he is well grounded in science.

Section 5.4 Other Sciences

How's the glass? Twenty-nine, I see.

—Sherlock Holmes, *The Boscombe Valley Mystery*

ASTRONOMY

We get our initial glance of Sherlock Holmes as astronomer in the very first story, STUD. It is in STUD that Watson makes his famous assessment of Holmes. Part of it reads:

Knowledge of Astronomy—Nil

Holmes appears to be unaware of how the solar system works. And when Watson describes it to him, Holmes vows to forget it because it won't help him solve crimes. In the early stories, we're dealing with the super-practical Holmes. He is interested only in things that have direct application to his work. Who cares about the solar system?

However, by the time of the forty-second story, BRUC, things have changed. He is shocked to receive a telegram from brother Mycroft announcing his imminent arrival at Baker Street. No longer ignorant of the solar system, Holmes states that for the famously lazy Mycroft to leave the comfort of the Diogenes Club to come to the Baker Street lodgings is as likely as a planet leaving its orbit. But there were clues that Holmes had gotten up to speed in astronomy long before this.

Our first hint appears in *The Musgrave Ritual* (MUSG), the twentieth story. He figures out the correct position of the sun for his calculation of where the shadow of the oak tree will fall (see section 5.1). He also notes that Brunton's intelligence is "quite first-rate." Therefore Holmes feels that he will not need to take into account "the personal equation[6] as the astronomers have dubbed it." He is saying that Brunton will not have made any errors. The point here is that Holmes is now referring to astronomers, indicating that he has been doing some reading in the field.

We next hear Holmes mention astronomy in the twenty-fourth story, *The Greek Interpreter* (GREE). Watson and Holmes have a discussion on the "obliquity of the ecliptic." The earth's orbital plane around the sun is called the ecliptic plane. If the earth were not tilted and had its axis of rotation perfectly upright, the obliquity of the ecliptic would be zero (Ridpath 2006, 132). But earth currently has an axial tilt of about 23.5° away from vertical. It is this tilt of the axis that gives the earth its seasons. The value of the tilt varies over the years between a minimum near 22.5° and a maximum near 24.5°. It is this variation in the earth's tilt, called the obliquity of the ecliptic, that Holmes and Watson discuss in GREE. We know that when they first met in STUD, Watson knew more astronomy than Holmes. Perhaps then it was he who led the astronomical conversation in GREE. We're not told. Holmes at the least knew enough now that he could participate in such a conversation.

It is obvious that Conan Doyle was an extremely well-read individual. Even so, it is surprising that a physician/author would insert a comment in his writings about "the change in the obliquity of the ecliptic." Was Conan Doyle reading accounts of current astronomical research? It turns out that he had a personal friend in the astronomical community. Alfred Drayson lived near Conan Doyle in Southsea near Portsmouth in the 1880s. Drayson was actually a patient of Conan Doyle the physician (Schaefer 1993). Drayson and Conan Doyle vacationed

[6] The variations or errors in observation or judgment caused by individual characteristics.

together. Who is this close friend to whom Conan Doyle would later dedicate a book?

Alfred Drayson made a career in the military, graduating in 1846 from the Royal Military Academy in Woolwich. After military service in India, South Africa, and North America (Stashower 1999, 95), Drayson returned to Woolwich to be an instructor of astronomy at his alma mater. He also did some part-time work at the observatory at Greenwich. In 1868, he was elected to the Royal Astronomical Society. Conan Doyle was so impressed by Drayson that he considered him a genius (Booth 1997, 122) and compared him favorably to Copernicus (Stashower 1999, 95). In March of 1890, Conan Doyle published a collection of ten short stories under the title *The Captain of the Polestar*. He dedicated the book to Drayson (Booth 1997, 134):

> To my friend Major-General A. W. Drayson as a slight token of my admiration of his great and as yet unrecognized services to astronomy.

Drayson did publish the results of his astronomy researches, but some of his work did not stand the test of time. An 1875 paper of particular interest to Sherlock Holmes readers was entitled *Variation on the Obliquity of the Ecliptic* (Schaefer 1993, 176). It proposes a theory that proved to be wrong. He also gave a lecture in 1884 to the Portsmouth Literary and Scientific Society on "The Earth and its Movement." In the lecture, he described the obliquity of the ecliptic (Booth 1997, 98). Conan Doyle was a member of the society and very likely went to hear his friend speak. Drayson's 1888 book, *Thirty Thousand Years of the Earth's Past History*, discusses variations in the obliquity of the ecliptic. It is nearly certain that Conan Doyle got the idea to use the obliquity of the ecliptic from his friend Alfred Drayson. In Section 5.1, we saw that the level of conversation between the roommates was on a high mathematical level. Again this is true as they talk about astronomy.

The other major astronomical topic in the Canon involves that other astronomer, Professor Moriarty. We're told in *The Final Problem* (FINA) that his "Treatise on the Binomial Theorem" had secured a chair in mathematics for Moriarty. But his most impressive work was astronomical. His *The Dynamics of an Asteroid* was a "book which ascends to such rarefied heights of pure mathematics" that few could even read it. So the professor had gravitated to astronomy once he became a faculty member. Even after moving on to become London's crime lord, Moriarty retained an interest and expertise in astronomy. When Inspector MacDonald goes to Moriarty's study to question him, the professor can't resist explaining eclipses to the inspector. He even gives a demonstration of how eclipses occur. He concludes by lending MacDonald a book on the topic (VALL).

But Moriarty's major effort in astronomy dealt with asteroids or "minor planets." Since the 1700s, astronomers have had an equation for computing the distances of the planets from the sun. It is called the Titius-Bode Law:

> D (in A.U.) = $0.4 + (0.3 \times N)$
> Where N = 0, 1, 2, 4, 8, etc. (doubling)
> D is in astronomical units (The distance of the earth from the sun is defined as 1 astronomical unit.)

This equation gives good estimates for the actual distances, as shown below (Kowal 1996, 2).

Planet	N	Calc. D	Measured D.
Mercury	0	0.4	0.39
Venus	1	0.7	0.72
Earth	2	1.0	1.00
Mars	4	1.6	1.52
GAP	8	2.8	2.77 (Ceres)
Jupiter	16	5.2	5.20
Saturn	32	10.0	9.54

One notable feature of these calculations is the gap between Mars and Jupiter. The existence of this gap caused astronomers to search for a missing planet. What was found instead was the first asteroid. It was in Sicily in 1801 that Giuseppe Piazzi discovered Ceres at 2.77 A.U. He named it to honor the patron goddess of Sicily (Kowal 1996, 1). Note how closely its distance from the sun matches the 2.80 value in the list above. The discovery of Ceres was soon followed by that of Pallas in 1802. This second asteroid was named in honor of Pallas Athena, the Greek goddess of wisdom. Eventually hundreds of asteroids would be found in the "asteroid belt" between Mars and Jupiter.[7]

These discoveries caused great excitement in the scientific world. Soon there were theories explaining why asteroids were found instead of another planet. Chemists took note of these astronomical advances by naming the next two chemical elements to be discovered after these two asteroids. Cerium and palladium were found in 1803. By the time of Conan Doyle and Holmes, excitement about asteroids had died down because hundreds were known by then. But in 1898, the first "near-earth" asteroid, Eros, was discovered. Never does an astronomical thing or event play a significant role in one of Holmes's cases. The most interesting thing

[7] Ceres is no longer considered an asteroid. In 2006, when Pluto was downgraded from planet to dwarf planet, Ceres was upgraded from asteroid to dwarf planet. There are currently officially five dwarf planets.

about the astronomy in the tales is how it got there. Conan Doyle's familiarity with Alfred Drayson's work on the obliquity of the ecliptic and the continuing interest in asteroids brought about the astronomical references in the Canon. Conan Doyle was able to emphasize the scientific literacy of Holmes and Watson by making them knowledgeable about the current state of astronomy.

GEOLOGY

Watson's original assessment of Holmes the geologist, given in STUD, is "practical, but limited." When he tried to recall the geology rating in *The Five Orange Pips* (FIVE), he misremembers just as he did with chemistry. Instead of "practical," Watson now says "profound." Was Holmes's knowledge of geology practical or profound? There's no way to know since, unlike chemistry, there is very little geology in the sixty stories. In both STUD and FIVE, Watson's analysis focuses on Holmes's ability to identify soils and connect them with areas of London, and perhaps beyond. This is hardly profound geology.

There are several instances in which Holmes does make use of this skill. In STUD, Watson reports that Holmes

> [a]fter walks has shown me splashes upon his trousers, and told me by their colour and consistence in what part of London he had received them.

In SIGN, Holmes applies this knowledge to reddish soil on Watson's shoe. He is able to state that Watson has been to the Wigmore Street Post Office. Holmes knows that the pavement has been removed exposing the reddish soil and that it is hard to avoid when entering the building. In FIVE, Holmes makes a similar deduction about a client. He deduces that John Openshaw has come up to London from the southwest.

> "You have come up from the south-west, I see."
>
> "Yes, from Horsham."
>
> "That clay and chalk mixture which I see upon your toe caps is quite distinctive."

These three examples of Holmes's deductions about soils and localities are entertaining. But they neither further the stories much, nor are they significant geology. In fact, Holmes's deduction about Openshaw and Horsham has been disputed (Klinger 2005, 137).

One instance of soil providing a clue that helps identify a culprit occurs in *The Three Students* (3STU). An examination for a lucrative scholarship is scheduled. But the day before the test, Hilton Soames, a tutor and lecturer at the college, discovers that one of the candidates has snuck into his chambers and read the exam. No footprints or fingerprints are found at the scene. Holmes turns to the two pieces of black clay that were found in Soames's room. He notes that the clay has traces of sawdust on it. He is already suspicious of Gilchrist. Only he is tall enough to have looked into the tutor's window to see the exam papers on the desk. Gilchrist competes in the long jump and is the only athlete among the three students. Holmes is up at six the next morning to visit the athletic grounds where he finds sawdust-covered black clay in the long-jump pit.

Another instance of the use of soil in a case happens in *The Devil's Foot* (DEVI). As usual, only Holmes notices the soil on the windowsill at Mortimer Tregennis's house:

> The gravel upon the window-sill was, of course, the starting-point of my research.

When he discovers that the gravel is found only near Dr. Leon Sterndale's cottage, Holmes has his man. Confronted with this and other evidence, Sterndale confesses to the murder. But here again we have a case where Holmes considers that Sterndale had good reason to avenge the murder of his beloved Brenda Tregennis by her brother Mortimer. He tells Sterndale, a lion-hunter and African explorer, that he is free to return to Africa to continue his work.

There is one other topic that falls into the area of geology. In *The Engineer's Thumb* (ENGR), fuller's earth plays a role in the plot. It is a type of clay that had industrial uses in Holmes's time. It continues to have applications today. Since the 1960s, the major uses of fuller's earth have been to absorb oil and grease and as cat litter (Hosterman and Patterson 1992, 3). Fuller's earth takes its name from its former principal use, which was cleaning or "fulling" wool (Hosterman and Patterson 1992, 2). In Victorian London, it was mainly used as an agent to remove oil (lanolin) from wool, which could then be made into valuable cloth.

In ENGR, a counterfeiting gang has set up shop in the village of Eyford. They are using a powerful press in their coining operation. When it begins to malfunction, they persuade an engineer, Victor Hatherly, to come one evening to repair it. Their cover story is that they own land that has deposits of fuller's earth. They need secrecy so that they can purchase adjacent land after they convince investors that their operation will be a success. So they blindfold Victor Hatherly and take him on what he estimates is a twelve- mile carriage ride to the house where the press is located. Hatherly fixes a leaky cylinder, but then makes the mistake of saying he knows that the press is not being used to compress fuller's earth. One of the crooks, Col. Lysander Stark, locks Hatherly in the room with the press and turns it on. Hatherly escapes but not before his thumb is cut off by the machine. He is brought for treatment to Dr. Watson, who notifies Holmes. Holmes makes a brilliant deduction about the location of the house by asking Hatherly about the condition of the horse when it arrived to take him there. Holmes is the only one to realize that a fresh horse had not come twelve miles to fetch the engineer. The twelve-mile ride was merely six miles away from the station, and then six miles back. The counterfeiter's house was right near the Eyford train station. But they escape before Holmes arrives in Eyford, never to be apprehended.

METEOROLOGY

We close the chapter with a discussion of the most surprising scientific topic of all. In BOSC, Holmes has been summoned by Inspector Lestrade to help with a murder case in the west of England. As they ride the train westward, Holmes relates what he knows of the case to Watson. He also remarks that the train is traveling "fifty miles an hour." In SILV, Holmes calculates a train speed of 53.5 miles per hour. There he tells us that "the calculation is a simple one" and explains how he did it (see section 5.1). The remark about train speed in BOSC has attracted little interest, perhaps because it appears to be more of an estimate.

When they arrive at Herefordshire, Inspector Lestrade has a carriage ready to take Holmes to the crime scene. Surprisingly Holmes declines the offer. Usually he wants to examine the scene before others alter it. Recall how he complained in STUD about the "herd of buffaloes" that obliterated much of the footprint evidence. Also since this murder was committed outside, it would seem even more urgent to go to the scene at once. If it was to start raining, for example, the crime-scene data may become compromised. But Holmes is confident that no rain is on the way, and thus there is no need for haste to go to Boscombe Pool. How does he know that it will not rain? Holmes checks the barometer (i.e., the "glass"). The age-old principle that rain accompanies low pressure is, presumably, his guide.

Mercury barometers made their appearance in the mid-1600s. But since October 2009, the sale of new ones has been banned in the United Kingdom. The elemental liquid mercury that fills the inverted tube in such barometers is now considered too toxic. Old mercury barometers can be restored, and individuals can construct their own. Needless to say, barometers were very much more common in 1890s England than they are now (Rothman 1990, 137):

> A barometer was usual in the hall of every middle-class English home.

Barometers may have been viewed as attractive accoutrements in homes. Watson considers the barometer in the hall at Mrs. Cecil Forrester's house to be an indication of a "tranquil English home." He is glad that Mary Morstan, his future wife, is lodging there during the events surrounding the Agra treasure in SIGN. As Klinger reports (2006, 284), "Mercury wheel, stick, and marine barometers, beautiful glass-and-wood objects used to predict the weather and now prized as antiques, were often found in Victorian homes." This familiarity suggests that, in England then, just about everyone could tell a low barometric pressure from a high one. Was Sherlock Holmes an exception?

In a mercury barometer, the pressure exerted by the earth's atmosphere is sufficient to hold up a column of mercury that is 29.92 inches high. That is a normal value at sea level. At higher elevations, the barometric reading will be lower than that average. It will also vary a little locally as pressure fronts come and go. When Holmes sees that the value is 29 inches, he is assured that bad weather is not coming. Later he says,

> The glass still keeps very high. It is of importance that it should not rain before we are able to go over the ground.

Incredibly he retires for the night without ever seeing the ground around the murder site. Twenty-nine inches is a very, very low value. It is a strong indicator of stormy weather. But Holmes's luck holds because next day "the morning broke bright and cloudless." As we saw (section 3.3), Holmes solves this mystery using the still-intact, undisturbed footprints.

Who should we blame for the bad science here? Was Sherlock Holmes ignorant when it came to meteorology? Was Arthur Conan Doyle? Shall we accept that ingenious explanation offered by Schweichert (1980, 244) that the barometric pressure was so low that Holmes's (and everyone else's) perceptions were altered, leading to the misstatement? Sherlockians have a tendency never to blame Holmes. They might very well attribute the remark to an error by Watson as he wrote up the case for publication. As with all Holmesian issues, you are free to form your own opinion.

Conclusion

When you have eliminated the impossible, whatever remains,
however improbable, must be the truth.
—Sherlock Holmes, *The Sign of the Four*

Sherlock Holmes and Professor Moriarty plunged over the Reichenbach Falls in *The Final Problem* (FINA), the twenty-sixth story (see figure C.1). What we read about the post-Reichenbach Holmes is that he had "never been the same man afterwards" (Stashower 1999, 443). Actually the very first story written after Holmes and Moriarty went over the falls was *The Hound of the Baskervilles* (HOUN). It is the most famous Holmes tale, and it is always rated as the very best one too. The next three stories, *The Empty House* (EMPT), *The Norwood Builder* (NORW), and *The Dancing Men* (DANC) are all rated fairly well. So Conan Doyle gets to the halfway point quite strongly (DANC is the thirtieth story). But soon the quality drops off. The fifty-six Holmes short stories have been rated several times (Bigelow 1993, 130–138). It is revealing to compare the first thirty stories with the last thirty. Here are the results from the 1959 ratings done by readers of *The Baker Street Journal.*

Figure C.1 Sherlock Holmes and Professor Moriarty on the precipice by the Reichenbach Falls in Switzerland

Rankings of the Sherlock Holmes Short Stories

Ten Best		*Ten Worst*	
Name	Story #	Name	Story #
SPEC	10	MAZA	49
REDH	4	VEIL	59
BLUE	9	YELL	17
SILV	15	BLAN	56
SCAN	3	3GAB	55
MUSG	20	CREE	51
BRUC	42	RETI	58
SIXN	35	LION	57
DANC	30	SUSS	52
EMPT	28	MISS	38

Eight of the ten stories on the "best" list are from the first half of the Canon. Only two later stories make the list. The "worst" list is just the reverse. Nine of the ten stories are from the second half; eight of the tales on the worst list are from the last twelve stories that Conan Doyle wrote, between 1921 and 1927. Even Conan Doyle himself agreed with this. In 1927, he listed his twelve favorite short stories and later added his next seven. Conan Doyle's list has fifteen early stories and four late ones.

> Arthur Conan Doyle's Favorite Holmes Short Stories
> SPEC (10), REDH (4), DANC (30), FINA (26), SCAN (3), EMPT (28), FIVE (7), SECO (40), DEVI (43), PRIO (32), MUSG (20), REIG (21), SILV (15), BRUC (42), CROO (22), TWIS (8), GREE (24), RESI (23), NAVA (25)

When the four long stories are included, not much changes. Generally HOUN displaces *The Speckled Band* (SPEC) as number one. But the later stories still fare poorly. One of the second-half tales that is always rated high is *The Bruce Partington Plans* (BRUC). This may be due to the "Mycroft Effect." An appearance by the ever-popular lazy brother gives BRUC added appeal. In addition, this story, as discussed in section 5.3, shows Holmes relying a good deal on his knowledge of science as he applies the physics of momentum and friction to help solve a murder.

In section 4.6, we mentioned Arthur Conan Doyle's shift to spiritualism. More than one literary critic has been gratified to note that Conan Doyle kept spiritualism and the occult out of his Sherlock Holmes work. But Conan Doyle was now less of a physician and man of science, and so was Holmes. Conan Doyle spent more and more of his time and energy in the spiritualist cause. So we find that in the latter half of the Holmes oeuvre, Conan Doyle began to leave science out. In chapter 4, we pointed out how Holmes drifted away from chemistry. There is little mention of it in the second half of the Canon. Tracy (1977, 70) lists seven stories (COPP, DANC, IDEN, NAVA, RESI, SIGN, and STUD) in which Holmes does chemical experiments. Every single one of these tales is from the first half of the Canon. In chapter 5, we saw a similar thing with regard to biology and mathematics. With regard to physics, we found that Holmes continued to use the magnifying lens throughout his career, but he did so most effectively early on. Of the references to astronomy, geology, and meteorology, fully 80 percent occur in the first half of the Canon. Holmes's use of scientific methods to solve his cases, discussed in chapter 3, also declined in the later stories, although not as dramatically. About 60 percent of the use of forensic science is in the first half of the Holmes Canon. The diminished presence of science in the later stories is obvious.

It is surely no coincidence that the stories that are short on science are generally viewed as inferior. Even Arthur Conan Doyle himself was well aware of this. He would often draw a laugh from banquet audiences by telling a story that made this very point (Higham 1976, 216):

> A Cornish fisherman was the worst critic I ever had. He told me, "Well sir, Sherlock Holmes may not have killed himself falling over that cliff. But he did injure himself something terrible. He's never been the same since!"[1]

There seems to be a cause-and-effect relationship between the use of science and the quality of the stories. When Holmes was portrayed as a detective actively using science in his work and life, the stories were full of appeal to readers. Science lent a robustness and complexity to the stories that contributed to their authenticity and provoked thought in the readers. In fact, it was Conan Doyle's idea from the start that a consulting detective who divined solutions in the absence of science and the scientific method would stretch even the simplest credulity. But one who applied the scientific method actively would challenge readers' faculties and impress everyone with a resourcefulness that, although occasionally improbable, was never impossible.

We turn to Isaac Asimov for a final thought. Previously we attempted to refute Asimov's criticisms of Holmes the chemist. But another of Asimov's articles on Holmes (1987, 204) got it just right. In this age of "Terminators" and special effects, we revere Sherlock Holmes because he is "someone who *thinks* rather than bashes."

[1] The exact wording varies depending on the source. See Lellenberg et al. 2007, 517.

APPENDIX

Doyle Scams

Holmes makes a remark about amber in *The Yellow Face* (YELL), but the wording in the American text differs from that in the English version. The English text reads:

> I wonder how many real amber mouthpieces there are in London? Some people think that a fly in it is a sign. Why, it is quite a branch of trade, the putting of sham flies into sham amber.

In the American text, the last sentence about sham amber is omitted. The result is an incomplete thought that leaves the reader wondering of just what the fly is a sign. Amber is a fossilized tree resin that can contain things, like flies, that were trapped in the substance millions of years ago (Klinger 2005, vol. 1, 451). The Natural History Museum of Great Britain has over 2,500 specimens of insects trapped in real amber (Kaye 1995, 299). But in Conan Doyle's time, there were plenty of unscrupulous people who created fake amber and put something in it to resemble a fly. This was done in an attempt to persuade the unwary that they were buying something ancient.[1] When chemists found a way to make synthetic resins in the 1940s, there was a surge in amber forgeries (Hoffmann 1990).

Conan Doyle shows here an awareness of scientific fakery that has led some to consider him as a perpetrator of other scientific frauds. The most sensational of these charges is that Sir Arthur Conan Doyle was the originator of the most famous fraud in the history of science, "Piltdown Man." This charge was made in the science news journal *Science 83* (Winslow and Meyer 1983). The charge was repeated and expanded in 1996 (Anderson 1996). Sherlockian scholars

[1] Such trickery came to mind when paleogeneticists failed in attempts to extract DNA from fossilized insects preserved in amber (*The New Yorker*, August 15 & 22, 2011, 67).

have reacted with outrage that the gentlemanly Conan Doyle should be labeled a fraud (Elliott and Pilot 1996).

In December 1912, Charles Dawson and Arthur Woodward announced the discovery of important fossils near the village of Piltdown in southern England. Piltdown Man seemed to be the perfect "missing link" in that he had a cranium that was humanlike and a jaw that was apelike. It should be noted that the articular condyle, the hinge of the jaw, was missing. Conspiracy theorists noted the similarity between "Arthur Conan Doyle" and "articular condyle," a bone that is distinctively different in apes and humans. In 1915, a second set of artifacts was found about two miles away at the Piltdown II site. By this point, Piltdown Man had taken his place in the anthropological evolutionary chain leading to homo sapiens.

However, as additional fossils were discovered around the world, they were consistent with one another and different from the Piltdown bones. All "intermediate forms" had a jaw that was humanlike and a cranium that was apelike. Piltdown Man had just the opposite. Not until 1949 was the telltale fluorine content of the Piltdown bones measured. Soon nitrogen analyses also demonstrated problems with the Piltdown fossils. A pigment known as van Dyke brown had been used. Now it was noticed that the teeth had been ground down to give the desired appearance. Striations were observed.

Piltdown Man was a hoax. The jaw was that of a juvenile female orangutan, about 500 or 600 years old, from the East Indies. Other animal bones were from the eastern Mediterranean area. The skull fragments were human. The teeth had been artificially filed. The articular condyle had been deliberately removed.

The premise of the *Science 83* article (Winslow and Meyer 1983) was that Conan Doyle had planted the bones in order to deliberately fool the scientific community. His purpose was to demonstrate that one fraud did not disprove all of science, and neither, then, should one fraudulent "medium" disprove all of spiritualism. The authors pointed out that Conan Doyle had visited all of the areas from which the bones were assembled; he lived within walking distance of the Piltdown site; and he was even photographed there. He had the chemical knowledge to do the fake staining of the bones. To the conspiracy theorists, he seemed like a perfect candidate. Much of the latter part of his life was spent promoting spiritualism. He spent a great deal of money and gave hours to the cause. He believed that his wife Jean was a medium. He worked diligently to persuade the magician Harry Houdini that spiritualism was real. If he could convince Houdini, then much of the world might also accept his claims. Richard Milner has been described as the "principal proponent of the Doyle theory" (www.tiac/net/~cri_a/piltdown/piltdown.html). He claims that the reason Conan Doyle didn't admit the hoax was that World War I was approaching, and he wished to be an advisor to the English government. Sir Arthur didn't think a scientific hoaxer would be welcome in that role (Kalush and Sloman 2006, 391).

I presented a poster paper[2] at the national meeting of the American Association for the Advancement of Science in San Francisco in January 1989. The poster mounted next to mine had the title "Doyle Scams." Being an admirer of Conan Doyle, I was quite interested in the evidence for the three scams discussed by the author, Charles L. Scamahorn of Berkeley, California. His main evidence that Conan Doyle was the Piltdown Man hoaxer was the above-mentioned similarity of his name to the missing articular condyle. Add to this the photo of Conan Doyle at Piltdown, and what more does a conspiracy lover need?

But there was more. Scamahorn also claimed that Conan Doyle planted the Kensington Runestone—between the towns of Holmes and Kensington in Minnesota. The evidence this time is that Conan Doyle visited the area in 1894, four years before the stone was discovered. Scamahorn claims that Conan Doyle, by making some fancy rearrangements of letters on the stone, had again inserted clues and was clearly spoofing the scientific community. How Conan Doyle transported the 202-pound stone is not explained. How he managed to entwine it in the roots of a poplar tree is also not explained. The Kensington Runestone is generally viewed as a fraud, having been denounced by several academics. Runic experts cite the style and type of the runes. Believers display it in the Runestone Museum in Alexandria, MN. They claim it demonstrates that Vikings made it to Minnesota in 1362.

Finally, Scamahorn claims that Conan Doyle also planted the Drake Plate near San Francisco when he was there in 1923. The "proof" is just as in the Kensington case: Conan Doyle visited the area in 1923. In addition, fanciful rearrangements of letters on the plate appear to Scamahorn to represent Conan Doyle's name. He is convinced that Conan Doyle planted it and left clues that others missed. In 1628, Francis Fletcher, the chaplain aboard Sir Francis Drake's ship *The Golden Hind*, wrote that Drake had placed a brass plate in the San Francisco Bay area in 1579. In 1936, such a plate was found. But its metallic composition was modern, 35.0 percent zinc and 64.6 percent copper (Lambert 1997, 194). Testing showed that the method of fabrication was "rolling," a modern process not available in Drake's time (Kaye 1995, 309). Someone planted the Drake Plate. Scamahorn is convinced that it was Conan Doyle.

Despite difficulty in getting his ideas accepted, Charles Scamahorn continues to push his theories about Conan Doyle. See the blog site probaway.wordpress.com, where in March 2009, he described the three scams. Then in December 2009 and January 2010, he presented his "proof" that Arthur Conan Doyle was the world-famous killer known as Jack the Ripper—enough said![3]

[2] My paper was entitled "The Calomel Rebellion." It had nothing to do with Sherlock Holmes or Arthur Conan Doyle.

[3] I still possess an autographed copy of the paper, inscribed to me as the first to see the "Doyle Scams."

References

Ackroyd, P. 2011. *London Under*. New York: Nan A. Talese/Doubleday.

Anderson, P. 1989. "A Treatise on the Binomial Theorem" in *Sherlock Holmes by Gas-Lamp*, P. Shreffler, ed. New York: Fordham University Press.

Anderson, R. B. 1996. "The Case of the Missing Link," *Pacific Discovery*, Spring Issue, 15–20.

Asimov, I. 1980. "The Problem of the Blundering Chemist," *Science Digest*, vol. 88(2), 8–17.

Asimov, I. 1987. "Thoughts and Sherlock Holmes," *The Baker Street Journal*, 37(4), 201–204.

Baring-Gould, W. S. 1967. *The Annotated Sherlock Holmes*. New York: Clarkson N. Potter.

Barthel, T. 2010. *Abner Doubleday: A Civil War Biography*, Jefferson, NC: McFarland.

Behn, N. 1994. *Lindbergh*. New York: The Atlantic Monthly Press.

Bengtsson, H. 1989. "And the Calculation is a Simple One," *The Baker Street Journal*, 39(4), 232–236.

Berdan, M. S. 2000. "The Ones That Got Away," *The Baker Street Journal*, 50(3), 23–30.

Bigelow, S. T. 1967. "Fingerprints and Sherlock Holmes," *The Baker Street Journal*, 17(3), 131–135.

Bigelow, S. T. 1961. "The Blue Enigma," *The Baker Street Journal*, 11(4), 203–214.

Bigelow, S. T. 1993. *The Baker Street Briefs*. Toronto: The Metropolitan Reference Library.

Bilger, B. 2012. "Beware of the Dogs," *The New Yorker*, February 27, 46–57.

Blank, E. W. 1947 "Was Sherlock Holmes a Mineralogist?," *Rocks and Minerals*, 22(3), 237.

Blinkhorn, S. F. 1993. "The Writing is on the Wall," *Nature*, 366, 208.

Blum, D. 2011. *The Poisoner's Handbook*. New York: The Penguin Press.

Born, W. 1937. "Purple," *Ciba Review*, 2, 106–117.

Booth, M. 1997. *The Doctor and the Detective*. New York: Thomas Dunne Books.

Bunson, M. E. 1994. *Encyclopedia Sherlockiana*. New York: Macmillan.

Burhoe, B. A. 2007. *Royal Canadian Mounted Police Dogs: The German Shepherd*. www.Goarticles.com.

Butler, W. S., and L. D. Keeney. 2001. *Secret Messages*. New York: Simon & Schuster.

Campbell, M. 1983. *Sherlock Holmes and Dr. Watson, A Medical Digression*. New York: Magico Magazine.

Caplan, R. M. 1989. "Why Coal-Tar Derivatives in Montpellier?," *The Baker Street Journal*, 39(1), 29–33.

Capuzzo, M. 2010. *The Murder Room*. New York: Gotham Books.

Cargill, A. 1890. "Health and Handwriting," *Edinburgh Medical Journal*, vol. 35, 627–631.

Carr, J. D. 1949. *The Life of Sir Arthur Conan Doyle*. New York: Vintage Books.

Cho, A. 2002. "Fingerprinting Doesn't Hold Up as Science in Court," *Science*, 295, January 18, 418.

Christ, J. F. 1947. *An Irregular Chronology of Sherlock Holmes of Baker Street*. New York: Magico Magazine.

Clark, J. D. 1964. "A Chemist's View of Canonical Chemistry," *The Baker Street Journal*, 14(3), 153.

Cole, S. A. 2001. *Suspect Identities: A History of Fingerprinting and Criminal Identification.* Cambridge, MA: Harvard University Press.

Cooke, C. 2005. "Mrs. Hudson: A Legend in Her Own Lodging-House," *The Baker Street Journal,* 55(2), 13–23.

Cooper, C. 2008. *Forensic Science.* New York: DK Publishing.

Cooper, P. 1976. "Holmesian Chemistry," in *Beyond Baker Street,* by M. Harrison . Indianapolis: Bobbs-Merrill.

Coppola, J. A. 1995. "A Chemist's View of Canonical Chemistry," *The Baker Street Journal,* 45(2), 106–113.

Coren, M. 1995. *Conan Doyle.* London: Bloomsbury.

Cox, M. 1993. *Victorian Detective Stories.* Oxford, UK: Oxford University Press.

Craighill, S. 2010. *The Influence of Duality and Poe's Notion of the Bi-part Soul on the Genesis of Detective Fiction in the Nineteenth Century,* thesis, Edinburgh Napier University.

Crump, N. 1952. *Sherlock Holmes Journal,* vol. 1(1), 16–23.

Curjel, H. 1978. "Death by Anoxia," *The Baker Street Journal,* 28(3), 152–156.

Dirda, M. 2012. *On Conan Doyle.* Princeton, NJ: Princeton University Press.

Douglas, J., and M. Olshaker. 1996. *Unabomber: On the Trail of America's Most-Wanted Serial Killer.* New York: Pocket Books.

Dove, G. N. 1997. *The Reader and the Detective Story.* Bowling Green, OH: Bowling Green University Popular Press.

Doyle, S., and D. A. Crowder. 2010. *Sherlock Holmes for Dummies.* Hoboken, NJ: Wiley Publishing.

Drayson, A. W. 1888. *Thirty Thousand Years of the Earth's Past History.* London: Chapman and Hall.

Duyfhuizen, B. 1993. "The Case of Sherlock Holmes and Jane Eyre," *The Baker Street Journal,* 43(3), 135–145.

Edwards, O. D., ed. 1993. *The Oxford Sherlock Holmes.* London: Oxford University Press.

Elliott, D., and R. Pilot. 1996. "Skull-Diggery at Piltdown," *The Baker Street Journal,* 46(4), 13–28.

Ellison, C. O. 1983. "The Chemical Corner," in *Sherlock Holmes and His Creator,* by T. H. Hall. New York: St. Martin's Press.

Faye, L. 2010. "Clay Before Bricks: Sherlock Holmes, Film Noir, and the Origins of the Hard-Boiled Detective," *The Baker Street Journal,* 60(3), 15–22.

Fetherston, S. 2006. "Shoscombe Through the Looking-Glass," *The Baker Street Journal,* 56(1), 41–50.

Fido, M. 1998. *The World of Sherlock Holmes.* Holbrook, MA: Adams Media Corporation.

Fincher, J. 1989. "Turning Bad Fingerprints Into Good Clues," *Smithsonian Magazine,* 20(7), 201.

Fisher, D. 1995. *Hard Evidence.* New York: Dell Publishing.

Fowler, A. 1994. "Sherlock Holmes and the Adventure of the Dancing Men and Women," in *Arthur Conan Doyle: Sherlock Holmes—The Major Stories with Contemporary Critical Essays,* J. A. Hodgson, ed. New York: St. Martin's Press.

Freese, P. L. 1986. "Howard Hughes and Melvin Dummar: Forensic Science Facts Versus Film Fiction," *Journal of Forensic Science,* 31(1), 342–359.

Garfield, S. 2001. *Mauve.* New York: W. W. Norton

Gerritsen, R., and R. Haak. 2007. *K9 Working Breeds.* Calgary, Canada: Detselig Enterprises LTD.

Gibson, J. M., and R. L. Green, eds. 1986. *Letters to the Press: Arthur Conan Doyle.* Iowa City, IA: University of Iowa Press.

Gillard, R. D. 1976. "Sherlock Holmes—Chemist," *Education in Chemistry,* 13, 10–11.

Graham, R. P. 1945. "Sherlock Holmes: Analytical Chemist," *Journal of Chemical Education,* 22, 508–510.

Green, R. L. 1983. *The Uncollected Sherlock Holmes.* London: Penguin Books.

Green, R. L. 1987. "The Evolution of Sherlock Holmes," *Baker Street Miscellanea*, no. 49, 2–9.
Green, R. L. 1990. "The Sign of the Four Or, The Problem of the Sholtos," *Baker Street Miscellanea*, no. 61, 1–3.
Greenwood, N. N., and A. Earnshaw. 1984. *Chemistry of the Elements*. Oxford, UK: Pergamon Press.
Haining, P., ed. 1995. *The Final Adventures of Sherlock Holmes*. New York: Barnes and Noble Books.
Hammett, L. P., and F. A. Lowenheim. 1934. "Electrolytic Conductance by Proton Jumps: The Transference Number of Barium Bisulfate in the Solvent Sulfuric Acid," *Journal of the American Chemical Society*, 56, 2620.
Harris, J. J. 1986. "The Document Evidence and Some Other Observations About the Howard R. Hughes 'Mormon Will' Contest," *Journal of Forensic Science*, 31(1), 365–375.
Higham, C. 1976. *The Adventures of Conan Doyle*. New York: Pocket Books.
Hiss, T. 1999. *The View From Alger's Window*. New York: Alfred A. Knopf.
Hodgson, J. A., ed. 1994. *Sherlock Holmes: The Major Stories with Contemporary Critical Essays*. Boston: Bedford Books of St. Martin's Press.
Hoffmann, R. 1990. "Blue as the Sea," *American Scientist*, 78, 308–309.
Holroyd, J. E. 1959. *Baker Street By-ways*. New York: Otto Penzler Books.
Holstein, L. S. 1954. "7. Knowledge of Chemistry—Profound," *The Baker Street Journal*, 4(1), 44–49.
Hosterman, J. W., and S. H. Patterson. 1992. *Bentonite and Fuller's Earth Resources of the U.S.*, U. S. Geological Survey Professional Paper 1522. Washington, DC: U.S. Government Printing Office.
Huber, C. L. 1987. "The Sherlock Holmes Blood Test: The Solution to a Century-Old Mystery," *The Baker Street Journal*, 37(4), 215–220.
Hudson, R. L. 1994. "Scotland Yard Stalks Printers' Prints," *The Wall Street Journal*, October 13, B1.
Hunt, H. 2011. "The Blue Carbuncle: A Possible Identification," *The Baker Street Journal*, 61(3), 45–48.
Inman, C. G. 1987. "Sherlockian Distillates," *Journal of Chemical Education*, 64(12), 1014–1015.
Jackson, J. 2009. *Using Chihuahuas in Police Work*. www.articlesbase.com.
Jacoby, S. 2009. *Alger Hiss and the Battle for History*. New Haven, CT: Yale University Press.
Jann, R. 1995. *The Adventures of Sherlock Holmes: Detecting Social Order*. New York: Twayne Press.
Jones, H. E. 1994. "The Origin of Sherlock Holmes," in *The Game is Afoot*, M. Kaye, ed. New York: St. Martin's Press.
Jones, P. K. 2011. "The Untold Tales Itemized," *The Baker Street Journal*, 61(2), 15–25.
Kalush, W., and L. Sloman. 2006. *The Secret Life of Houdini*. New York: Atria Books.
Kasson, P. 1961. "The True Blue," *The Baker Street Journal*, 11(4), 200–202.
Kaye, B. H. 1995. *Science and the Detective*. New York: VCH Publishers.
Kellogg, R. L. 1989. "Watson's Psychoanalytical Touch," *Baker Street Miscellanea*, no. 59, 44–45.
Kendall, J., and A. W. Davidson. 1921. "Compound Formation and Solubility in Systems of the Type Sulfuric Acid: Metal Sulfate," *Journal of the American Chemical Society*, 43, 979–990.
King, L. R., and L. S. Klinger. 2011. *The Grand Game*. New York: The Baker Street Irregulars.
Klinger, L. S. 2011. "Some Trifling Observations on "The Dancing Men," *The Baker Street Journal*, 61(4), 23–24.
Klinger, L. S., ed. 2005 & 2006. *The New Annotated Sherlock Holmes, vols. I, II, and III*. New York: W. W. Norton & Co.
Koppenhover, K. M. 2007. *Forensic Document Examination*. Totowa, NJ: Humana Press.
Kowal, C. T. 1996. *Asteroids: Their Nature and Utilization*, second edition. West Sussex, UK: Praxis Publishing Ltd.
Lachtman, H. 1985. *Sherlock Slept Here*. Santa Barbara, CA: Capra Press.
Lambert, J. B. 1997. *Traces of the Past*. Cambridge, MA: Perseus Publishing.

Lane, B. 2005. *Crime and Detection*. New York: DK Publishing.

Leavitt, R. K. 1940. "Nummi in Arca," in *Studies in Sherlock Holmes*, V. Starrett, ed. New York: Otto Penzler Books.

Lellenberg, J., D. Stashower, and C. Foley. 2007. *Arthur Conan Doyle: A Life in Letters*. New York: Penguin Press.

Liebow, E. 1982. *Dr. Joe Bell: Model for Sherlock Holmes*. Bowling Green, OH: Bowling Green University Popular Press.

Macintyre, B. 1997. *The Napoleon of Crime, The Life and Times of Adam Worth, Master Thief*. New York: Broadway Paperbacks.

Matlins, A. L., and A. C. Bonanno. 1989. *Gem Identification Made Easy*. Woodstock, VT: GemStone Press.

Matlins, A. L., and A. C. Bonanno. 1993. *Jewelry and Gems: The Buying Guide*. Woodstock, VT: GemStone Press.

McGowan, R. J. 1987. "Sherlock Holmes and Forensic Chemistry," *The Baker Street Journal*, 37(1), 10–14.

McKinney, C. E. 2011. *Indigo*. New York: Bloomsbury.

McSherry, F. D., M. H. Greenberg, and C. G. Waugh, eds. 1989. *The Best Horror Stories of Arthur Conan Doyle*. Chicago: Academy Chicago Publishers.

Michell, J. H., and H. Michell. 1946. "Sherlock Holmes the Chemist," *The Baker Street Journal*, 1(3), 245–252.

Miller, R. 2008. *The Adventures of Arthur Conan Doyle*. New York: St. Martin's Press.

Moenssens, A. A., J. E. Starrs, C. E. Henderson, and F. E. Inbau. 1995. *Scientific Evidence in Criminal and Civil Cases*, fourth edition. Westbury, NY: The Foundation Press.

Moss, R. A. 1982. "A Research into Coal-Tar Derivatives," *The Baker Street Journal*, 32(1), 40–42.

Moss, R. A. 1991. "Brains and Attics," *The Baker Street Journal*, 41(2), 93–95.

Moss, R. A. 2011. "Arthur Conan Doyle and Sherlock Holmes A Philatelic Celebration," *American Philatelist*, 125(8), 736–742.

Murphy, B. F. 1999. *The Encyclopedia of Murder and Mystery*. New York: Palgrave.

Murray, E. science.marshall.edu/murraye/Footprint%20Lab.html.

Musto, D. F. 1968. "A Study in Cocaine: Sherlock Holmes and Sigmund Freud," *Journal of the American Medical Association*, 204(1), 125–132.

Musto, D. F. 1988. "Why Did Sherlock Holmes Use Cocaine?," *The Baker Street Journal*, 38(4), 215–216.

Mutrux, H. 1977. *Sherlock Holmes: Roi des Tricheurs*. Paris: Pensée Universelle.

Nez, C. 2011. *Code Talker*. New York: The Berkley Publishing Group.

Ozden, H., Y. Balci, C. Demirustu, A. Turgut, and M. Ertugrul. 2005. "Stature and Sex Estimate Using Foot and Shoe Dimensions," *Forensic Science International*, 147, 181–184.

Paige, R. 2002. *Death at Dartmoor*. New York: The Berkley Publishing Group

Park, O. 1994. *The Sherlock Holmes Encyclopedia*. New York: Carol Publishing Group.

Phillips, D. P., et al. 2001. "The Hound of the Baskervilles Effect: Natural Experiment on the Influence of Psychological Stress on the Timing of Death," *British Medical Journal*, 323 (7327), 1443–1446.

Pratte, P. 1992. "Cocaine and the Victorian Detective," *The Baker Street Journal*, 42(2), 85–88.

Priestman, M. 1994. "Sherlock Holmes—The Series," in *Arthur Conan Doyle: Sherlock Holmes—The Major Stories with Contemporary Critical Essays*, J. A. Hodgson, ed. New York: St. Martin's Press.

Propp, W. W. 1978. "A Study in Similarity," *The Baker Street Journal*, 28(1), 32–35.

Putney, C. R., J. A. Cutshall King, and S. Sugarman. 1996. *Sherlock Holmes Victorian Sleuth to Modern Hero*. London: The Scarecrow Press.

Puttnam, C. 1991. "Science: Can Police Dogs Really Sniff Out Criminals?," *New Scientist*, September 14, 24.

Rafaeli, A., and R. J. Klimoski. 1983. "Predicting Sales Success Through Handwriting Analysis: An Evaluation of the Effects of Training and Handwriting Sample Content," *Journal of Applied Psychology*, 68(3), 212–217.

Redmond, C. 1981. "In Praise of the Boscombe Valley Mystery," *The Baker Street Journal*, 31(3), 170–174.
Redmond, C. 1993. *A Sherlock Holmes Handbook*. Toronto: Simon and Pierre.
Redmond, D. A. 1982. *Sherlock Holmes: A Study in Sources*. Montreal: McGill Queens University Press.
Redmond, D. A. 1964. "Some Chemical Problems in the Canon," *The Baker Street Journal*, 14(3), 145–152.
Rendall, V. 1934. "The Limitations of Sherlock Holmes," in *Baker Street Studies*, H. W. Bell, ed. New York: Otto Penzler Books.
Rennison, N. 2005. *Sherlock Holmes: The Unauthorized Biography*. New York: Grove Press.
Ridpath, I. 2006. *Astronomy*. New York: DK Publishing.
Riley, D., and P. McAllister. 1999. *The Bedside, Bathtub & Armchair Companion to Sherlock Holmes*. New York: Continuum Publishing.
Robbins, L. M. 1985. *Footprints*. Springfield, IL: Charles C. Thomas.
Roberts, R. M. 1989. *Serendipity: Accidental Discoveries in Science*. New York: John Wiley & Sons.
Rothman, S., ed. 1990. *The Standard Doyle Company, Christopher Morley on Sherlock Holmes*. New York: Fordham University Press.
Rutland, E. H. 1974. *An Introduction to the World's Gemstones*. New York: Doubleday.
Saferstein, R. 1995. *Criminalistics: An Introduction to Forensic Science*. Englewood Cliffs, NJ: Prentice Hall.
Saltzman, M. D., and A. L. Kessler. 1991. "The Rise and Decline of the British Dyestuffs Industry," *Bulletin for the History of Chemistry*, 9, 7–15.
Sartain, J. S. 2008. "Surgeon General William A. Hammond (1828–1900): Successes and Failures of Medical Leadership," *Gunderson Lutheran Medical Journal*, 5(1), 21–28.
Sayers, D. 1929. *An Omnibus of Crime*. Garden City, NJ: Garden City Publishing.
Schaefer, B. E. 1993. "Sherlock Holmes and Some Astronomical Connections," *The Baker Street Journal*, 43(3), 171–178.
Schmidle, N. 2011. "Getting Bin Laden," *The New Yorker*, August 8, 34–45.
Schmidt, F. L., and J. E. Hunter. 1998. "The Validity and Utility of Selection Methods in Personnel Psychology: Practical and Theoretical Implications of 85 Years of Research Findings," *Psychological Bulletin*, 124, 262–274.
Scholten, P., M.D. 1988. "The Connoisseurship of Sherlock Holmes with Observations on the Place of Brandy in Victorian Medical Therapeutics," *Baker Street Miscellanea*, no. 54, 1–7.
Schweickert, W. 1980. "A Question of Barometric Pressure," *The Baker Street Journal*, 30(4), 243–244.
Shreffler, P., ed. 1989. *Sherlock Holmes by Gas-Lamp*. New York: Fordham University Press.
Silvermann, K. 1991. *Edgar A. Poe*. New York: Harper Perrennial.
Simpson, H. 1934. "Medical Career and Capabilities of Dr. J. H. Watson," in *Baker Street Studies*, H. W. Bell, ed. New York: Otto Penzler Books.
Simpson, K. 1983. *Sherlock Holmes on Medicine and Science*. New York: Magico Magazine.
Sinkankas, J. 1962. *Gem Cutting—A Lapidary's Manuel*. New York: Van Nostrand Reinhold.
Smith, D. 2009. *The Sherlock Holmes Companion*. New York: Castle Books.
Sova, D. B. 2001. *Edgar Allan Poe: A to Z*. New York: Checkmark Books.
Specter, M. 2002. "*Do Fingerprints Lie?*," *The New Yorker*, May 27, 96–105.
Starrett, V. 1930. *The Private Life of Sherlock Holmes*. New York: Macmillan.
Starrett, V. 1934. "The Singular Adventures of Martha Hudson," in *Baker Street Studies*, H. W. Bell, ed. New York: Otto Penzler Books.
Stashower, D. 1999. *Teller of Tales: The Life of Arthur Conan Doyle*. New York: Henry Holt and Co.
Stinson, R. 2003. "Art in the Aniline Dye," *The Baker Street Journal*, 53(1), 25–27.
Sullivan, M. R. 1996. *Sherlock Holmes: Victorian Sleuth to Modern Hero*, C. R. Putney, J. A. Cutshall King and S. Sugarman, eds. Lanham, MD: Scarecrow Press.
Swift, W., and F. Swift. 1999. "The Associates of Sherlock Holmes," *The Baker Street Journal*, 49(1), 25–45.

Symons, J. 1979. *Portrait of an Artist: Conan Doyle*. London: Whizzard Press.

Tansey, R. G., and F. S. Kleiner. 1996. *Gardner's Art Through the Ages*, tenth edition. Fort Worth, TX: Harcourt Brace College Publishers.

Tracy, J. 1977. *The Ultimate Sherlock Holmes Encyclopedia*. New York: Gramercy Books.

Travis, A. S. 2007. "Mauve and Its Anniversaries," *Bulletin for the History of Chemistry*, 32(1), 35–44.

Trenner, N. R., and H. A.Taylor. 1931. "The Solubility of Barium Sulphate in Sulphuric Acid," *Journal of Physical Chemistry*, 35, 1336–1344.

Utechin, N. 2010. "From Piff-Pouff to Backnecke: Ronald Knox and 100 Years of Studies in the Literature of Sherlock Holmes," *The Baker Street Journal 2010 Christmas Journal*.

Vail, W. A. 1996. "Premature Burial: New in the Annals of Crime?," *The Baker Street Journal*, 46(3), 7–12.

Vatza, E. J. 1987. "An Analysis of the Tracing of Footsteps from Sherlock Holmes to the Present," *The Baker Street Journal*, 37(1), 16–21.

Wagner, E. J. 2006. *The Science of Sherlock Holmes*. Hoboken, NJ: John Wiley & Sons.

Walters, L. R. 1978. "The Hydrocarbon Puzzle," *The Baker Street Journal*, 28(4), 222–223.

Waterhouse, W. C. 2004. "What Was the Blue Carbuncle?," *The Baker Street Journal*, 54(4), 19–21.

Welcher, F. J. 1957. "History of Qualitative Analysis," *Journal of Chemical Education*, 34(8), 389–391.

White, G. E. 2004. *Alger Hiss's Looking-Glass Wars*. Oxford, UK: Oxford University Press.

Winslow, J. H., and A. Meyer. 1983. "The Perpetrator at Piltdown," *Science 83*, September, 33–43.

Wislicenus, J. 1885. *Adolph Strecker's Short Textbook of Organic Chemistry*, second edition. London: Keoan Paul, Trench, & Co.

Zerwick, P. 2011. www.phoebezerwick.com.

Index